Integrated Security Science

Edited by
Genserik Reniers, Nima Khakzad, Pieter Van Gelder

Volume 1

Reniers, Khakzad, Van Gelder (Eds.)
Security Risk Assessment
Integrated Security Science

Security Risk Assessment

In the Chemical and Process Industry

Edited by
Genserik Reniers, Nima Khakzad, Pieter Van Gelder

DE GRUYTER

Editors
Genserik Reniers
TU Delft
Safety&Security Science Group
Building 31
Jaffalaan 5
2628 BX Delft
The Netherlands

Nima Khakzad
TU Delft
Safety&Security Science Group
Building 31
Jaffalaan 5
2628 BX Delft
The Netherlands

Pieter Van Gelder
TU Delft
Safety&Security Science Group
Building 31
Jaffalaan 5
2628 BX Delft
The Netherlands

ISBN 978-3-11-050052-3
e-ISBN (PDF) 978-3-11-049908-7
e-ISBN (EPUB) 978-3-11-049776-2
Set-ISBN 978-3-11-049909-4
ISSN 2367-0223

Library of Congress Cataloging-in-Publication Data
A CIP catalog record for this book has been applied for at the Library of Congress.

Bibliographic information published by the Deutsche Nationalbibliothek
The Deutsche Nationalbibliothek lists this publication in the Deutsche Nationalbibliografie; detailed bibliographic data are available on the Internet at
http://dnb.dnb.de.

Typesetting: le-tex publishing services GmbH, Leipzig
Printing and binding: CPI books GmbH, Leck
Cover image: ollo/E+/gettyimages
♾ Printed on acid-free paper
Printed in Germany

www.degruyter.com

Preface

Responsible organizations in the chemical industry understand that they are facing a wide variety of threats that may cause harm from intentional acts against them due to the nature and importance of the chemicals they manufacture, distribute, use, and store. While the industry has a long history of managing accidental risks, the efforts toward a similar level of security risk management has been more limited and is just now being more appreciated and understood.

There is an urgency in dealing with security in the chemical industry given the potential for such events and the possible harm that the chemicals may cause. Paramount to proper risk management is the recognition of and evaluation of security risks to identify consequences, vulnerabilities, and threats in an organized, systematic, and repeatable fashion. This information is invaluable to inform management of means to reduce these risks to acceptable levels.

Security risk assessment (SRA) is the foundation of the security risk management (SRM) system since one cannot make optimal decisions when uninformed of security risks. To be effective, SRM is a formal management system on a par with safety, environmental, quality, and other business objectives, and the organization integrates security risk considerations into an enterprise risk management system. It is required to treat the analysis of security in a dynamic process since the threats may rapidly change.

The terrorist acts of 2001 on the United States and many global events since then have created a wave of concern for chemical security since it is a potential means to create large-scale asymmetrical attacks. At the time the industry did not have commonly accepted or standardized best practices for conducting SRAs. It was recognized by governments and professional societies in the industry that development and adoption of SRA methods was required, and considerable guidance has been developed provide information about the best practices.

SRAs done effectively blend with accidental risk assessments to find optimal solutions to achieve higher levels of performance in chemical safety and security. Organizations that embrace SRA and SRM as instrumental practices in their firms will benefit greatly from better decision making on design and operation of facilities and their businesses. It is possible to improve security with intelligent analysis and a holistic consideration of risk in nontraditional ways, such as through inherent security, isolation, defense in depth, and operational risk management rather than only through added layers of security.

With the prevalence of chemicals, the global development of the chemical industry including facilities and the value chain, and the escalating global tensions, the world is once again seeing abuse of standard industrial chemicals for nefarious purposes. These events are stark reminders that require governments and private companies responsible for manufacturing and handling chemicals to find ways to reduce risks while maintaining a vibrant industrial base.

David A. Moore, PE, CSP
President & CEO AcuTech Group, Inc.
www.acutech-consulting.com

https://doi.org/10.1515/9783110499087-201

Contents

List of Contributors

Francesca Argenti
Università di Bologna
LISES - Dipartimento di Ingegneria Civile,
Chimica, Ambientale e dei Materiali
Via Terracini n. 28
40131 Bologna
Italy
francesca.argenti@unibo.it

Shailendra Bajpai
National Institute of Technology
Department of Chemical Engineering
Jalandhar
India
bajpais@nitj.ac.in

Paul Baybutt
Primatech Inc.
50 Northwoods Blvd.
Columbus, OH 43235
USA
paulb@primatech.com

Valerio Cozzani
Università di Bologna
LISES - Dipartimento di Ingegneria Civile,
Chimica, Ambientale e dei Materiali
Via Terracini n. 28
40131 Bologna
Italy
valerio.cozzani@unibo.it

Pieter Van Gelder
Delft University of Technology
Safety and Security Science Group
Jaffalaan 5
2628 BX Delft
The Netherlands
p.h.a.j.m.vangelder@tudelft.nl

J.P. Gupta
Shiv Nadar University
Dadri (UP)
India
jpg@snu.edu.in

Cecilia Haskins
NTNU - Norwegian University of Science
and Technology
Department of Mechanical
and Industrial Engineering
Trondheim
Norway
cecilia.haskins@ntnu.no

Nima Khakzad
Delft University of Technology
Safety and Security Science Group
Jaffalaan 5
2628 BX Delft
The Netherlands
n.khakzadrostami@tudelft.nl

Gabriele Landucci
Dipartimento di Ingegneria Civile e Industriale,
Università di Pisa, Largo Lucio Lazzarino n. 2,
56126 Pisa
Italy
gabriele.landucci@unipi.it

Nicola Paltrinieri
NTNU - Norwegian University of Science
and Technology
Department of Mechanical
and Industrial Engineering
Trondheim
Norway
nicola.paltrinieri@ntnu.no

Hans J. Pasman
Texas A&M University
Mary Kay O'Connor Process Safety Center
(MKOPSC)
College Station, TX 77843
USA
hjpasman@gmail.com

Genserik Reniers
University of Antwerp
Engineering Management
Faculty of Applied Economic Science
Prinsstraat 13
2000 Antwerp
Belgium
genserik.reniers@uantwerpen.be

Luca Talarico
University of Antwerp
Engineering Management
Faculty of Applied Economic Science
Prinsstraat 13
2000 Antwerp
Belgium
luca.talarico@uantwerpen.be

Laobing Zhang
Delft University of Technology
Safety and Security Science Group
Jaffalaan 5
2628 BX Delft
The Netherlands
laobing.zhang@tudelft.nl

1 Introduction

Safety and security both concern the avoidance and mitigation of losses of different origins (safety looks at possibly unintentionally caused losses, while security is about tackling deliberately caused losses). Our society today is ever more focused on security. However, in the past decades, mainly safety issues were tackled and predominantly safety research was carried out to obtain continuous improvement and higher health and safety figures in organizations of any kind. Questions such as “what is safety?”, “what is risk?”, “how to manage safety adequately?”, “who is responsible for safety? ”, “how safe is safe enough?”, and other complex topics arose and were investigated by practitioners around the world. They were also studied by researchers from insurance companies as well as by academics and research institutes. Moreover, politicians and regulators came into the discussion with ever more interest in the safety subject. This conglomerate of stakeholders has led to the huge progress that has been made with respect to safety in the past century.

In fact, according to Reniers and Khakzad [2] two safety revolutions took place: (i) the “safety first movement” (1900s until 1950s) represents the first safety revolution, and (ii) the “risk management and loss prevention” approaches (1960s until 2010s) denote the era of the second safety revolution. The third safety revolution to further advance safety, especially in the chemical industry, is summarized by the acronym CHESS (from 2020s onwards). Figure 1.1 shows the three safety revolutions along with the underlying theories, models, concepts and ideas per decade.

Reniers and Khakzad [2] indicate that the third safety revolution can be represented by the acronym CHESS. CHESS, in fact, summarizes five very important fields where revolutionary progress is needed:
- **C**luster-thinking and intensified cooperation
- **H**igh transparency and efficient inspection
- **E**ducation, training and learning
- **S**ecurity integration
- **S**afety innovation and dynamic risk assessments

At first sight, these fields represent a well-known recipe for improving safety in any industry whereas they are nothing new. However, one should realize that the combination of these domains could indeed lead to a third safety revolution in the chemical industry if they were addressed in radical innovative ways. The required innovation can be exemplified by a number of concrete ideas, which can only be realized if the current mentality of practitioners, academics and people from the authorities changes. For more information, we refer to Reniers and Khakzad [2]. With respect to security, the authors indicate that innovation is to be expected with respect to more effective counter-terrorism security practices in the chemical industry, especially in the era of blind and violent terrorism. At present, security efforts in chemical plants are aimed at addressing low-impact high-frequency security risks such as burglary

https://doi.org/10.1515/9783110499087-001

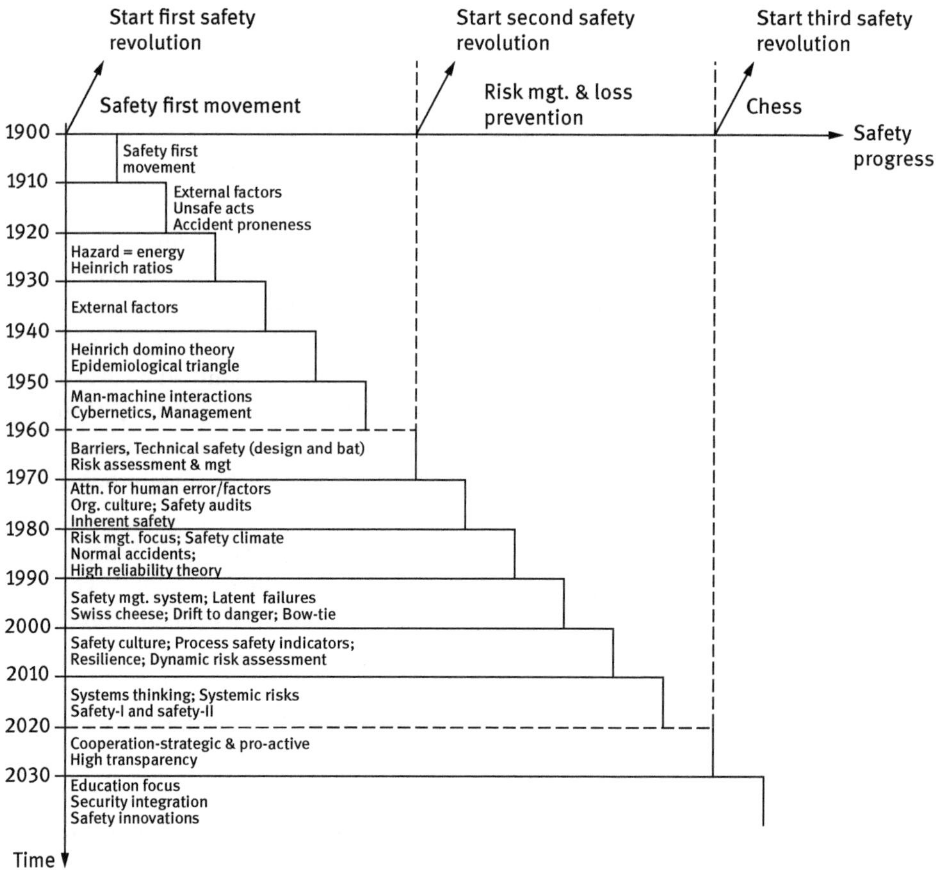

Fig. 1.1: Safety progress and the three safety revolutions (applicable to the chemical industry) (1900–2030 and future) [2].

and sabotage, or, at best, amateur terrorists. However, an adequate upgrade is needed to address high-impact low-frequency security events (anti-terrorist security measures) preferably from an inherent design based viewpoint [1]. However, more generally, security should be treated in an integrated way with safety by company safety management, thereby respecting the specificities accompanying security.

In fact, when one looks at a safety risk, three components are necessary: hazards, exposures (to the hazards) and losses. A hazard can be seen as a potential, a condition, a circumstance, a characteristic, or the like, that (with a certain likelihood) might cause losses. As such, hazards are characterized by the lack of deliberateness. Safety risks may thus lead to losses that were suffered without any human intention. In the case of safety risks, nature or random failures have caused the losses, or people have done their best to not cause losses, but nevertheless incidents/accidents/losses have occurred, or people have violated rules, but in their mind it was with the best intention, for instance to increase production, or to speed up a working process (and not to deliberately cause losses). Safety risk assessment is nothing more or less than the identification, analysis, evaluation and prioritization of all possible hazards, exposures and unintentional losses. Safety risk management is based on safety risk assessment and deals with treating/decreasing/lowering as many hazards, exposures and unintentional losses as possible, using a variety of management approaches and safety measures (so-called safety barriers or safety functions).

Security risks, if expressed analogously to safety risk, are also composed of three elements: threats, vulnerabilities (to the threats) and losses. Threats (comparable with "hazards" in safety) can be seen as possible individuals or groups of individuals possibly wanting to deliberately cause losses. As such, threats imply intention. Security risks thus result from deliberate human actions to cause losses, be it through physical attack or cyber-attack. Vulnerabilities (comparable with "exposures" in safety) are those weaknesses (in people, infrastructure, reputation, etc.) that make a critical asset susceptible to the threats. Security risk assessment is thus concerned with identifying, analyzing, evaluating and prioritizing possible threats, vulnerabilities and intentional losses. Security risk management is based on security risk assessment and handles the treating/decreasing/lowering as many threats, vulnerabilities and intentional losses as possible, employing security management techniques and security measures (so-called countermeasures).

It is obvious that the relatively new field of security can learn a lot from the past decades of research in the old field of safety. Hundreds of safety risk assessment methods exist, whereas only a very limited number of security risk assessments are known. Moreover, risk assessment in the safety domain has already made great progress with respect to quantification and probabilistic and dynamic thinking. In the security risk assessment domain, much has yet to be developed concerning quantification and probabilistic and dynamic thinking. There are undoubtedly knowledge spillovers from safety towards security that may be exploited. Nevertheless, there are also fundamental differences between the two fields that cannot be denied, for instance the transparency difference. Both the comparisons and the differences between the two fields of science will become much clearer when reading this book. Figure 1.2 illustrates how this book is constituted.

Chapter 1 explains the importance of the topic, especially in our current complex society and this era of ruthless terrorism. In Chapter 2, Pasman tackles a sensitive issue: physical security legislation and regulations for chemical plants, discussing and comparing the pros and cons of such legislation and directives in the United States versus those in Europe. A plea is made for a more centralized methodical security policy in Europe, when it concerns hazardous materials in chemical plants that may cause devastating disasters if they were to be misused by intelligent terrorists. Chapter 3 by Baybutt provides an overview of what security constitutes and how it should be seen within the range of potential disastrous events in a process plant. This

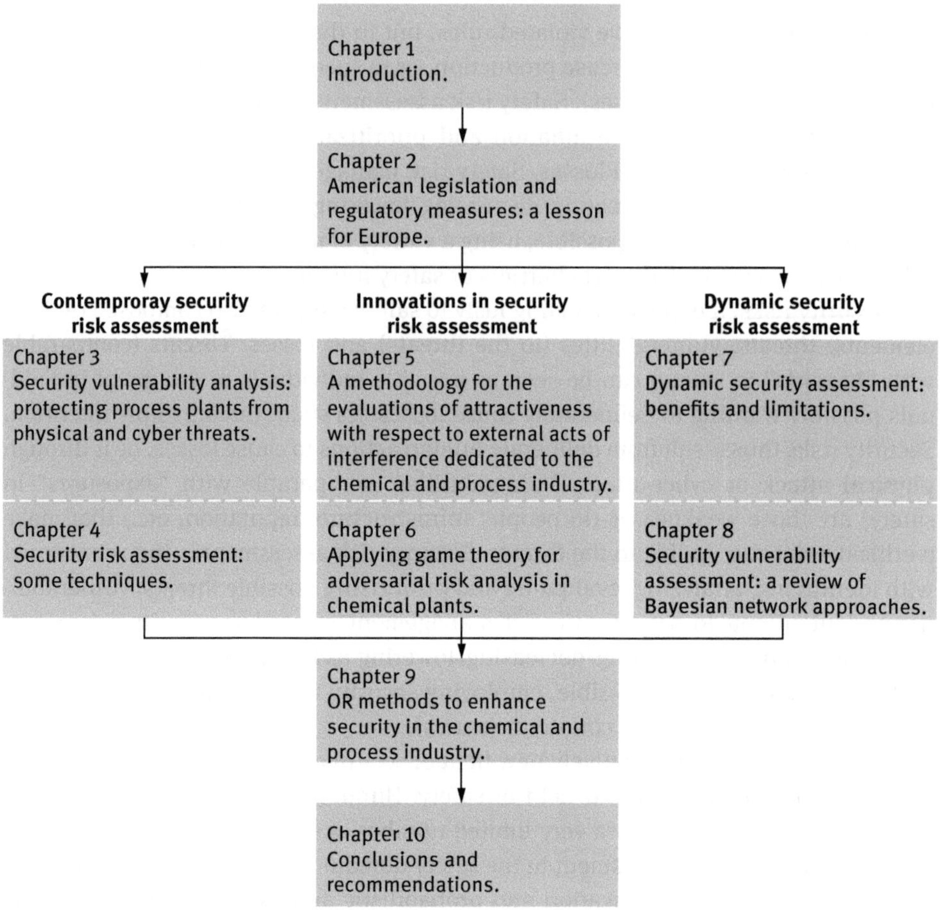

Fig. 1.2: Chapters of the book.

chapter also explains what the current insights in applying security risk assessments are, what terminology is used, how to interpret the results, etc. In Chapter 4, Bajpai and Gupta put forward a current approach of security risk assessment. Although there might be some overlap with Chapter 3, it does not pose a problem for the reader. Chapter 4 helps increase the legibility of security terminology and leads to an understanding of security risk assessment techniques currently used in the chemical industry. Chapter 5 by Landucci et al. elaborates on one of the factors of security risk assessment, that is, the "attractiveness" of a chemical plant from a security adversary perspective. This complicated issue is captured by the authors suggesting an innovative approach using the combination of a hazard based attractiveness index (quantitative assessment) and a site-specific so-called induction factor (qualitative estimate). Chapter 6 by Zhang and Reniers introduces game theory as a powerful mathematical tool into security applications in the chemical and process industry.

The authors argue that a security risk assessment, as it is used today in chemical facilities, needs an upgrade towards more reliability and optimality, especially in the allocation of security budget and resources. Game theory can help achieve much more rational, less "belly-feeling driven", decision making about physical security countermeasures. In Chapter 7, Paltrinieri and Haskins further expand on making the security assessment techniques more dynamic. Besides the potential of making static methods more dynamic, some advanced security assessment methods are discussed. In Chapter 8, Khakzad discusses the application of Bayesian network for making security risk assessment approaches more dynamic. Khakzad explains the advantages of using Bayesian networks in combination with utility theory and game theory, in the form of influence diagrams, to further advance probabilistic thinking in the discipline of security assessment. Chapter 9 concludes the list of contributions in this volume, indicating how methods commonly used in the field of operations research, can be employed to enhance security in the chemical and process industries. Talarico and Reniers categorize the models into four different sets of applications: mitigation, preparedness, response and recovery, depending on the lifecycle of an attack. In each of these groups, the decision making process can be supported by operations research models and methods. In Chapter 10, overall conclusions are drawn and recommendations based on the contributions are formulated.

References

[1] Reniers G, Amyotte P. Prevention in the chemical and process industries: future directions. J Loss Prevent Proc. 2012; 25(1):227–231.

[2] Reniers G, Khakzad N. Revolutionizing safety and security in the chemical and process industry: applying the CHESS concept. J Integ Sec Sci. 2017; 1(1):2–15.

Hans Pasman

2 American legislation and regulatory measures: a lesson for Europe?

2.1 Introduction to critical infrastructure protection and public safety and security

People as part of a society share common infrastructures to enable their activities, to prosper, and to maintain a satisfactory safety and security level. These infrastructures have over time become rather diverse. Very basic in any society today and taken by most people for granted are clean water supply and a sewer system. Further, electrical energy, and energy in the form of energy carriers such as at present natural gas and other fossil fuels, and in view of climate change, possibly in the future, hydrogen, are most important. Apart from the many sources of electricity such as fossil fuel combustion in power stations, nuclear and hydro power, and today the more sustainable solar power and wind energy, there are the very costly connecting energy transportation grids to equalize demand. Other essential demands besides food are communication and transportation, implying a host of wired and wireless grids, and an extensive road and rail system.

Most of the above is physical, but society is dependent on many organizational and economical networks, and infrastructural channels such as finance and banking, health, education, justice and police, and emergency response. There is also much interconnectivity and interdependency among networks. This implies the risk of propagation of breakdown from one network to another resulting in a crisis. As natural disasters not only lead to fatalities and injuries, or destroy housing and roads, and may cause large scale damage to such infrastructural networks as electricity and water supply, a country's economy may be severely damaged. Therefore, the UN and other international bodies encourage national governments to build resilience into critical infrastructure.

Within the scope of this book, this chapter on regulatory measures shall consider the protection of industrial physical infrastructure and not social or economic networks but production plants and transport systems. More specifically, we shall focus on the protection of the process industry encompassing primarily the energy and

---

Dr.Ir. Hans J. Pasman has been in various management positions in the Defense Research part of the TNO Applied Research Organization for more than 30 years. He is Emeritus Chemical Risk Management of the Delft University of Technology and currently Research Professor at the Mary Kay O'Connor Process Safety Center (MKOPSC) of the Texas A&M University, College Station, Texas, USA. The views presented in this chapter are his personal ones.

https://doi.org/10.1515/9783110499087-002

chemical industry but strictly speaking also steel making, food processing, and other industry that in their processing have large quantities of materials with hazardous properties. The security of nuclear power plants is a much older issue for which regulatory bodies and regulations have been in existence for many years and which will not be described here, as we shall focus on measures taken this century in view of the increased threats of terrorist attacks and extreme weather.

Perhaps the share of the chemical industry in the overall infrastructure is not always recognized by the public. Chemical industry produces a large variety of materials and substances that we need in daily life and for which no natural resources exist, either in quality or in quantity. One can think of fibers for clothing, building materials, coatings, fertilizers, and thousands of other products, also supporting the economy as a whole by producing materials used in further manufacturing processes. Therefore, part of this industry is considered to be critical infrastructure. Some parts of it can be considered to be of strategic interest, as these are unique sources of materials and substances that are crucial for maintaining economic and defense related activities.

Threats are diverse; they will range from extreme weather effects to intentional external attacks of facilities or sabotage. Terrorist attacks can have the goal of initiating a release of hazardous material that will threaten plant workers and the surrounding population to cause large scale economic damage or inhibit the use of certain materials, or to steal materials or substances to fabricate lethal weapons. A terrorist attacks can be by intrusion and access to facilities or by use of remote attack means from outside either from land, ship, or air. This can be physical by means of weaponry and explosive effects, or by using nuclear, biological, chemical, or radiological means (NBCR) against people, or by cyber-attack interfering with process control.

Setting up a major part of the infrastructural systems has been initiated, guided, and controlled by governments, and built and maintained by governmental agencies, in some cases after private initiatives already taken became nationalized. However, in particular in the 1990s within the background of a free global market philosophy of higher efficiency, ownership of much of all infrastructure shifted to private enterprise. As government remains accountable in a general sense for protection of infrastructure that is critically important to society, in the late 1990s and the early part of the 21st century the developed countries took initiatives in this respect. This trend was strengthened in view of the increasing complexity of the economy driven by higher demands and new technology, the larger interdependency of organizations and individuals and with that the increasing vulnerability of the well-being of society. Incidents such as disastrous electricity "blackouts", and attempts to intentionally inflict damage such as spreading highly contagious diseases, and others made it clear that something should be done.

So, in 1998 because of growing potential vulnerability, President Bill Clinton initiated an action to establish protection of critical infrastructure by issuing Presidential Directive PDD-63 [1]. This formed the start of a US National program on Critical Infrastructure Protection, abbreviated as CIP. After the terrorist attack on the Manhattan World Trade Center on September 11, 2001, efforts have been multiplied. By the Homeland Security Presidential Directives, of which the first one appeared on October 29, 2001 [2], the main organizational anti-terrorist lines were been drawn. This was followed with the Critical Infrastructures Protection Act of 2001 [3], while the Homeland Security Act of 2002 [4] founded the Department of Homeland Security (DHS) and made that department accountable for homeland security. Hurricane Katrina in 2005 and later hurricane disasters added to the necessity of having adequate preventive and protection measures and increased resilience by being prepared.

Europe followed the development in the US with some delay. In 2004 the European Commission was asked by the Council to develop a CIP strategy. A year later, on November 17, 2005 the Commission adopted the Green Paper [5] offering a number of policy options. Near the end of 2006, the Commission provided an update to the Council on its activities, entitled: On a European Programme for Critical Infrastructure Protection (EPCIP) [6]. Finally, 2 years later this resulted in Council Directive 2008/114/EC of December 8, 2008 [7] on the identification and designation of European critical infrastructures and the assessment of the need to improve their protection. The underlying problems to arrive at a common European policy and the final content of the directive will be treated in Section 2.3. For a more extensive summary of the history of CIP programs in the USA and in Europe, reference is made to the book of Alessandro Lazari [8].

The energy and IT/communications sectors have been always high on the priority list of items to be protected. Driven by the requirement to protect the manufacture of strategically important materials within one's own country, to protect the population around plants from violent phenomena in the case of an attack on a store of hazardous material, and the threat of theft of explosive material ingredients, chemical facilities have been also on the list.

In the next sections, we first shall take a more detailed look at the US regulation, which has been installed over the last decade with respect to the infrastructure sectors, and subsequently at the European regulatory measures.

2.2 US critical infrastructure protection actions

As mentioned above, after the initiative of President Clinton in 1998 to establish CIP [1] it took until after the attack on the Twin Towers in 2001 and the attempts to spread anthrax in 2002, for President Bush to sign the Homeland Security Act of 2002 [4] establishing the Department of Homeland Security (DHS). The act clearly served the purpose for federal, state, and local authorities and organizations of being better prepared and able to respond to terrorism and to take preventive action. In addition, it reinforced the critical infrastructure protection by bringing in the executive Federal Emergency Management Agency (FEMA), which had existed since 1978, and a number of other institutions with tasks in the case of major disaster under a newly formed Directorate of Emergency Preparedness and Response of the DHS.

Further, the Directorate of Border and Transportation Security was established under DHS. In addition to Customs Service, this directorate also contains the Transportation Security Administration (TSA). The latter is charged with the security of, in particular, airports, e. g., concerning explosives detection, in consultation with the Federal Aviation Agency (FAA). As we shall see later, a number of security tasks with respect to transportation of hazardous materials, which were under the wings of the Department of Transportation shifted to the TSA of DHS. Quite a number of other pre-existing entities, such as the US Coast Guard, were absorbed by the new department (22 altogether according to Bucci and Inserra [9]). In 2006, DHS issued its first National Infrastructure Protection Plan (NIPP) with updates appearing in 2009 and 2013 [10]. The plan provides general information on vision, mission, and goals, the CI environment, fundamentals, collaborations, and action call.

The establishment of such a new department also gave rise to criticism. The critics were concerned about the loss of individual freedom as a result of the powers of the new department and the large costs it would incur. From an organizational point of view not only the internal coordination of the various directorates would be a huge task but also establishing many new liaising ties with other departments. In any case, an important tie with respect to CIP in the physical sense is the one with the Department of Energy (DoE) of which the CIP activity we shall consider next. A further important physical security activity, which falls directly under the DHS, is implementation of the Chemical Facility Anti-Terrorism Standards (CFATS) Act [11] established in 2007 and tracking of how it works out. We shall describe the act and regulations and what it meant for the chemical industry in below.

However, before describing the organizations with a direct executive for CIP and the measures they took, it is relevant to note some observations made by President Obama in the 2013 Presidential Policy Directive – Critical Infrastructure Security and Resilience, PPD-21 [12]. This directive came out after a decade of experience with DHS. Obama's directive re-emphasized how crucial functioning of critical infrastructural assets, networks, and systems are to public confidence and to the US. The directive uses words, such as "vital" to "safety, prosperity, and well-being of the Nation." It acknowledged the complexity of the problem. It also clearly coupled security with *resilience* for the first time. The latter will make actions more complete, because besides preventing and protecting, also emergency response, contingency, and recovery are given more weight. CIP shall obtain more of a holistic system approach. CIP shall not be improvised activity but planned and provided with necessary resources to minimize damage. Strengthening of security and resilience shall be against both *physical* and *cyber* threats, and *all hazards* shall be considered. The directive underscored shared responsibility for CIP of federal, state, local, tribal, and territorial (e. g., Puerto Rico) entities and that of owners and operators of the infrastructure. It even

Tab. 2.1: Sectors of CIP and SSA.

	CIP Sector	Sector Specific Agency
1	Chemical	Department of Homeland Security
2	Commercial Facilities	Department of Homeland Security
3	Communications	Department of Homeland Security
4	Critical manufacturing	Department of Homeland Security
5	Dams	Department of Homeland Security
6	Defense Industrial Base	Department of Defense
7	Emergency Services	Department of Homeland Security
8	Energy	Department of Energy
9	Financial Services	Department of the Treasury
10	Food and Agriculture	Department of Agriculture and Department of Health and Human Services
11	Government Facilities	Department of Homeland Security and General Services Administration
12	Healthcare and Public Health	Department of Health and Human Services
13	Information Technology	Department of Homeland Security
14	Nuclear Reactors materials and Waste	Department of Homeland Security
15	Transportation Systems	Department of Homeland Security and Department of Transportation
16	Waste and Wastewater Systems	Environmental Protection Agency

announced engagement with international partners to strengthen their domestic CI if the US also depends on it. After stressing the interconnectedness and interdependency of the infrastructure, the directive identified energy and communications as the main enablers for all sectors.

The directive announced three strategic imperatives for the Federal Government clearly addressing weaknesses in the past: 1) refining and clarifying functional relationships across the government, 2) identification of baseline data and systems requirements, and 3) implementation of an integration and analysis function to inform planning and operations decisions. Secretary DHS leads the efforts strategically, and the Sector Specific Agencies (SSAs) conduct all that is needed. The roles and responsibilities of DHS as a department with respect to vulnerability assessments, situational awareness, and coordination of activities are spelled out. Moreover, the tasks of a number of other departments, commissions and agencies to provide information or contributing otherwise are described.

Then, the imperatives are explained in more detail, and attention is paid to innovation and research and development. Finally, an action plan with a time schedule is given to achieve the goals specified in the directive. For example, within 240 days the Secretary DHS shall demonstrate "near real-time situational awareness capability for critical infrastructure that includes threat streams and all-hazards information as well as vulnerabilities" DHS shall further provide the status of critical infrastructure and potential cascading effects, support decision making and disseminate all information needed to contain damage throughout an incident. Metrics shall be developed to measure abilities and risks and updated regularly when changes are observed. The CIP plan as a whole must be updated, including the functional relationships within DHS, across federal departments, and the public–private partnerships (PPS), introduced by President Clinton's PDD [1], evaluated, and if necessary improved.

The directive PPD-21 distinguishes 16 CIP sectors and their lead agencies. These are summarized in Table 2.1. Relevant for the remaining of this chapter on process industry are energy and chemicals with their respective SSAs, the Department of Energy, and the Department of Homeland Security. In both cases, hundreds of private companies own and operate the assets.

The Center for Infrastructure Protection and Homeland Security of the George Mason University, School of Business in Arlington, Virginia, is an example of a university that conducts comprehensive analyses and research to improve the safety and security of the United States and its allies across all critical infrastructure sectors. It is involved in CIP-related education and also issues the monthly CIP Report [13] with topics and news related to themes such as the energy sector or resilience.

2.2.1 Energy security and assurance

The nuclear part of energy, nuclear power stations with everything that belongs to them such as nuclear waste and waste processing, which emanated from nuclear weapons development, has been already been the subject of health, safety, security, and environmental regulations for many years. In this context, it has been subjected to oversight by commissions, e. g., the Nuclear Regulatory Commission, and inspection by independent authorities. The Department of Energy has been dealing with issues adhering to adequate risk control for many years. So, the increased vigilance with respect to possible terroristic attack arising at the beginning of this century can be considered as an enhancement and not as the completely new element it has been for the chemical industry, as we shall see in the next section.

Because of the many players in the field of power generation and delivery, both governmental and the private Department of Energy (DoE) have set up a special Office of Enterprise Assessment (EA) to carry out DoE's Independent Oversight Program to assure safety and security, see oversight implementation, DOE Order 226.1B [14] and its program, DOE Order 227.1A [15]. The program consists of conducting appraisals with the assistance of trusted agents. Naturally, EA shall be given all information requested and access to plants. The appraisals may contain force-on-force security exercise testing. Findings shall be documented, but imminent dangers or major vulnerabilities shall be notified immediately to the manager involved. EA is described in Independent Oversight Program Appraisal Process Protocols in more detail as well as how an appraisal is conducted [16]. The focus here is on nuclear installations and materials. EA reports to Congress [17] at the end of the year.

The DoE Office of Electricity Delivery and Energy Reliability (OE) is responsible for security and resilience of the grid. The Assistant Secretary for Electricity Delivery and Energy Reliability, Patricia Hoffman, in her testimony for the House in 2015 [18] gave an overview of current and expected challenges such as the diversity of green sources of electricity connected to the grid in addition to the conventional ones and the load by the possible future mass market of electric cars. With respect to the fiscal year 2016, she made a plea for investments in the grid with a priority on protection of the grid from all hazards, secondly to invest in transformer resilience, and thirdly in cyber security.

It can therefore be concluded that protection of critical energy infrastructure captures attention on a national level, is covered by regulation, and that there is a sincere political will to invest in the required strengthening.

2.2.2 Chemical facility anti-terrorism standards

A special policy regarding the chemical industry was not foreseen yet in the Homeland Security Act of 2002 [4]. It was only late in 2006 that Section 550 of the Homeland Security Appropriations Act gave DHS the authority to promulgate a special (interim final) rule concerning prevention of terroristic attack at facilities of the chemical industry, and this was signed in 2007. This rule became the Chemical Facility Anti-Terrorism Standards (CFATS) Act, of which the latest version was signed in 2014, see [19]. In fact, three possible terrorist/criminal operations must be thwarted. These are 1) intentionally causing disastrous mishap that would cause a major hazard to personnel and residential population; 2) disrupting a process installation so that the strategic supply of certain materials is interrupted or completely destroyed; 3) theft of materials that can be used to fabricate explosives, or highly toxic chemical agents. Such operations can be realized from outside by intrusion, or from inside by act of sabotage. The intrusion can be cyber wise: hacking controls or even taking over the operation remotely. It can also be physically carried out through release of hazardous material due to damaging/rupturing/ penetrating tanks, vessels, or piping by explosive blasts. The latter can also be realized by flying over a plant area, e. g., by drone.

Before we go into more detail about the contents of the CFATS and the organization to implement and maintain it, Presidential Executive order 13650 of August 1, 2013 [20] will be mentioned. This order was prompted by the disastrous ammonium nitrate (AN) detonation accident initiated by fire that killed 15 people, the majority fire fighters, and caused colossal damage in the town of West, in Texas on April 13, 2013. The accident was thoroughly investigated by the Chemical Safety Board [21]. The accident showed once again that hard prilled Fertilizer Grade AN (FGAN) can detonate in a fire; the 30 t portion of AN that detonated had been stored in combustible plywood bins inside a storage building containing 40–60 t AN in total, while a railcar present on the site contained another 100 t. (The railcar overturned but the cargo did not detonate.

Also, 17 t anhydrous ammonia was stored in pressure vessels and did not escape.) The West Fertilizer Company had not notified DHS of the amount of AN being stored. (The plant had many times more in store than the amount threshold for required notification, the so-called "top screen" submittal.) In 2016, the Federal Bureau of Investigation issued a notice seeking information on any information leading to the arrest of individuals involved in intentionally causing the West explosion, which confirmed that the investigation concluded that this was not an accidental event.

The executive order established a working group (WG) co-chaired by the Secretaries of DHS and Department of Labor and by the Administrator of The Environmental Protection Agency (EPA) and further consisting of representatives of various other departments and entities. The WG was ordered to accomplish within certain time periods a fair number of tasks with the goal to improve communication and collaboration with respect to safety and security between federal agencies, State regulators, and state, local, and tribal emergency responders, chemical facility owners and operators, and communities. The WG was further asked to consult the Chemical Safety and Hazard Investigation Board (CSB) whether existing working arrangements with and between Environmental Protection Agency (EPA), the Bureau of Alcohol, Tobacco, Firearms and Explosives (ATF), and the Occupational Health and Safety Administration (OSHA) should be improved. The WG was to consider whether modernization of policy, regulations, and standards is required, while EPA and OSHA was to scrutinize the two main safety regulations for the process industry, the Risk Management Program rule and the Process Safety Management rule, respectively. The WG was also to identify best risk management practices. This all shows how serious US leadership takes these matters. In May 2014, the group reported (in 121 pages) progress to the President [22] on actions already taken and future actions planned. Local emergency planning should be strengthened, the same holds for federal coordination and data exchange; OSHA sent out a Request for Information to stakeholders on what should be improved in the Process Safety Management standard, while EPA is training inspectors and planning modernization of the Risk Management Program rule. Finally, stakeholder feedback is being collected and best practices developed. The latter resulted in RAGAGEP (Recognized and Generally Accepted Good Engineering Practices) being enforced by OSHA.

Back to CFATS: the rule 6 CFR Part 27 of 2006 [22] established risk based performance standards. The rule explains to which type of facilities it applies and which form of a security risk. Chemical facilities that possess so-called chemicals of interest (COIs), being hazardous materials (substances) listed in Appendix A of the rule in a quantity over a certain threshold, must fill out and submit a top screen form. We shall come back to the procedure to be followed, but first some explanation of the standards is given. Congress did not give DHS the authority to specify any security measure – instead only to develop performance standards that must be met based on risk. In 2009, 3 years after the rule was published, a CFATS guidance document [23] came out focusing on the standards to be realized by owners and operators of chemical facilities. This document explains backgrounds, reasons, and modus operandi of CFATS, and assists a user in how to comply with the law. Fundamental is the meaning of a performance standard as applied in CFATS versus a design or technology-based standard. The former leaves a user free how to comply as long as the performance is achieved (goal oriented, and more cost-effective), whereas the latter prescribes exactly what must be done. There are 18 standards; these are summarized in Table 2.2. Each standard is explained in detail, while in tabular form metrics are shown with corresponding measures for each tier number of a facility, where, of course, the heaviest measures are to be applied for Tier 1. Details of the various possible measures are given in an appendix. The SSPs must include the measures taken to comply with the standards.

For following the application and plan submission procedures, DHS developed the online Chemical Security Assessment Tool (CSAT). This is a secure electronic, integrated system with which a user can provide the needed information online. Subsequently, DHS will assign the facility with one of four tiers, with tier 1 constituting the highest risk. An assigned and notified facility must conduct a security vulnerability assessment (SVA) and develop a site security plan (SSP). These two exercises are both performed by going online with CSAT and answering questions about the facility, while an instruction manual assists by

Tab. 2.2: Chemical Facility Anti-Terrorism Standards ([23], Section 27.230)

1	**Restrict Area Perimeter.** Secure and monitor the perimeter of the facility
2	**Secure Site Assets.** Secure and monitor restricted areas or potentially critical targets within the facility
3	**Screen and Control Access.** Control access to the facility and to restricted areas within the facility by screening and/or inspecting individuals and vehicles as they enter, including: (i) Measures to deter the unauthorized introduction of dangerous substances and devices that may facilitate an attack or actions having serious negative consequences for the population surrounding the facility and (ii) Measures implementing a regularly updated identification system that checks the identification of facility personnel and other persons seeking access to the facility and that discourages abuse through established disciplinary measures
4	**Deter, Detect, and Delay.** Deter, detect, and delay an attack, creating sufficient time between detection of an attack and the point at which the attack becomes successful, including measures to: (i) Deter vehicles from penetrating the facility perimeter, gaining unauthorized access to restricted areas or otherwise presenting a hazard to potentially critical targets (ii) Deter attacks through visible, professional, well-maintained security measures and systems, including security personnel, detection systems, barriers and barricades, and hardened or reduced-value targets (iii) Detect attacks at early stages, through countersurveillance, frustration of opportunity to observe potential targets, surveillance and sensing systems, and barriers and barricades and (iv) Delay an attack for a sufficient period of time to allow appropriate response through on-site security response, barriers and barricades, hardened targets, and well-coordinated response planning

Tab. 2.2: (continued) Chemical Facility Anti-Terrorism Standards ([23], Section 27.230)

5	**Shipping, Receipt, and Storage.** Secure and monitor the shipping, receipt, and storage of hazardous materials for the facility
6	**Theft and Diversion.** Deter theft or diversion of potentially dangerous chemicals
7	**Sabotage.** Deter insider sabotage
8	**Cyber.** Deter cyber sabotage, including by preventing unauthorized on-site or remote access to critical process controls, such as Supervisory Control and Data Acquisition (SCADA) systems, Distributed Control Systems (DCSs), Process Control Systems (PCSs), Industrial Control Systems (ICSs) critical business systems and other sensitive computerized systems
9	**Response.** Develop and exercise an emergency plan to respond to security incidents internally and with the assistance of local law enforcement and first responders
10	**Monitoring.** Maintain effective monitoring, communications, and warning systems, including: (i) Measures designed to ensure that security systems and equipment are in good working order and inspected, tested, calibrated, and otherwise maintained (ii) Measures designed to regularly test security systems, note deficiencies, correct for detected deficiencies, and record results so that they are available for inspection by the Department and (iii) Measures to allow the facility to promptly identify and respond to security system and equipment failures or malfunctions
11	**Training.** Ensure proper security training, exercises, and drills of facility personnel
12	**Personnel Surety.** Perform appropriate background checks on and ensure appropriate credentials for facility personnel, and, as appropriate, for unescorted visitors with access to restricted areas or critical assets, including: (i) Measures designed to verify and validate identity (ii) Measures designed to check criminal history (iii) Measures designed to verify and validate legal authorization to work and (iv) Measures designed to identify people with terrorist ties
13	**Elevated Threats.** Escalate the level of protective measures for periods of elevated threat
14	**Specific Threats, Vulnerabilities, or Risks.** Address specific threats, vulnerabilities, or risks identified by the Assistant Secretary for the particular facility at issue
15	**Reporting of Significant Security Incidents.** Report significant security incidents to the Department and to local law enforcement officials
16	**Significant Security Incidents and Suspicious Activities.** Identify, investigate, report, and maintain records of significant security incidents and suspicious activities in or near the site
17	**Officials and Organization.** Establish official(s) and an organization responsible for security and for compliance with these standards and
18	**Records. Maintain appropriate records.**

explaining how to use the tool, what data on the facility and its operations to provide, and suggesting options for answers. Before using CSAT SSP, the user must receive chemical terrorism vulnerability information (CVI) training. Based on the information provided, the tier number of the facility may be adapted. The required degree of thoroughness and detail of SVA and SSP will depend on the tier number.

Altogether, the process is described in four guidance documents published in the period 2008–2011 [24–27]. This was followed by an information brochure [28] for the sector in 2012, which provided many hints and clues to identify and recognize security threats. It contains a table of evacuation distances in the case of a vehicle borne improvised explosive device (VBIED), which might contain as much as 30 t explosives. It also gives suggestions to counter cyber threats, and how and what to report if suspicious activity is observed. Before the decision of compliance with the CFATS Act is made, inspections are held, while assistance by DHS is offered in the preparation stage.

In 2014, Dana A. Shea of the Congressional Research Service [29] reported about the state of play of the implementation of CFATS. Based on available information Shea concluded that DHS incurs an increasing backlog. The difference in the number of sites that have received a final tier assignment and have actually been inspected, and those that have had their site security plan authorized is increasing. A review of whether the plan complies with the standards can take a considerable amount of time and may require discussion with the owner or operator. The expectation is that eliminating this backlog will still take several years. An accurate prediction could not be given.

For industry the series application of registration, top screen, SVA and SSP led to long delays in determining their risk, and hence, to uncertainty on investment for possibly required security upgrade. In 2015, following critical remarks in Congress and industry about the long implementation delays, DHS issued the Expedited Approval Program for Tier 3 and 4 facilities [30] to reduce backlog. To further involve the chemical industry community DHS organizes annual chemical sector security summits and issues fact sheets with news and guidance. In 2006, DHS also established the Chemical Security Analysis Center (CSAC) to assess and identify vulnerabilities and respond to potential chemical threats and hazards. This center, located in Aberdeen, MD, initiated, e. g., the Jack Rabbit II chlorine dispersion trials to verify and validate models.

2.2.3 Maritime Transportation and Security Act 2002

About a year after the Twin Tower attack, the Maritime Transportation and Security Act 2002 [31] was promulgated. Obtaining this law was urgent because the US has many ports and large tonnages of goods and materials pass these ports. In addition, at the time, ferries transported 113 million people a year and 32 million vehicles. These numbers were growing. Also, the cruise line industry formed a risk. Ports are relatively open and vulnerable to terrorism, while an investigation a few years before revealed that this was also true for criminal activity such as smuggling and others. Moreover, as described at the end of this section, the International Maritime Organization meanwhile developed a new security system that should be followed. Responsibility for executing the law was attributed to the US Coast Guard, although at the time of promulgation it was not sure yet under which department it would function as the Department of Homeland Security was not yet founded. After DHS was established in the same year, the Coast Guard became DHS's military muscle.

The Act regulated the conduction of vulnerability assessments of ports and vessels, and the preparation of a National Maritime Transportation Security Plan under which Area Maritime Transportation Security Plans comprising security zones are developed. The Act introduced "avant la lettre" quite a few modes of operation in executing the law as later have been implemented in CFATS, such as requiring owners and operators of vessels and facilities to prepare and submit a Vessel and Facility Security Plan within a period of 6 months.

The detailed security regulation can be found in 33 Code of Federal Register, Chapter I, Sub-chapter H Maritime Security, Parts 101–107 [32] in which 103–106 describe Area, Vessels, Facilities and Outer Continental Shelf Facilities security, respectively. Part 128 treats security of passenger terminals.

In this context, the International Ship and Port Facility Security Code (ISPS Code) promulgated by the International Maritime Organization and coming into force in 2004 is also mentioned. This code serves to protect ships and port facilities against terrorist acts by specifying a set of minimum requirements, such as preventing unauthorized access to port restricted areas and vessels, and the unauthorized bringing in of weapons or explosives. This encompasses cooperation in security with respect to threat detection and prevention means, exchange of information, and assessment methods, as well as security plans and procedures. Development of the code was initiated after the attack on the New York WTC Twin Towers in September 2001, and the text was agreed upon by the 108 SOLAS (International Convention for the Safety of Life at Sea) signatories in 2002.

2.2.4 Transportation security of hazardous materials

Within DHS, the Transportation Security Administration (TSA) is among other tasks charged with the security of transportation of hazardous materials by any mode: road, rail, air, water, and pipeline. Coordination of DHS's TSA and the Department of Transportation (DOT) became institutionalized with the Homeland Security Act 2002 [4]. Homeland Security Presidential Directive No. 7, December 17, 2003, Critical Infrastructure Identification, Prioritization and Protection (HSPD-7) [33] ordered that DOT and DHS must collaborate in regulating the transportation of hazardous materials by all modes (including pipelines). This resulted in a Memorandum of Understanding (MOU) between DOT and DHS arranging how to coordinate, exchange information, and assist each other. For example, a 2006 Annex to this MOU between TSA and the Pipeline and Hazardous Materials Safety Administration (PHMSA) [34] detailed this collaboration further. Another MOU involving as a third partner the Nuclear Regulatory Committee is about the security of radioactive material transportation and was signed in 2015 [35].

In connection with security of hazardous materials transportation, further steps have been taken and a series of additions/amendments prepared to cover security aspects in the extensive existing safety rules of Title 49 CFR [36]. This encompasses Parts 172.802–822 on Safety and Security Plans, Parts 174 Carriage by Rail, Parts 175

Carriage by Aircraft, Parts 176 Carriage by Vessel, Parts 177 Carriage by Public Highway, Parts 190–196 Pipeline safety (Natural and other gases, LNG, oil, and hazardous liquids), and Parts 1520–1580 on the security of airport and related activities.

2.3 The EU critical infrastructure protection directive

The events in the US in 2001 were followed closely in Europe. In 2004, the European Council (Heads of Member States plus the President of the European Commission – EC) prepared a strategy for CIP. The EC organized two seminars with stakeholders, the second one including industry to ask for suggestions. This resulted in the 2005 Green Paper [37] setting out the main objectives of a European Programme for Critical Infrastructure Protection (EPCIP), and a supporting CIWIN, a Critical Infrastructure Warning Information Network). The paper also posed a number of questions of what the boundaries of the program should be in view of effectiveness and the common interest of an EPCIP on the EU level with regard to national CIPs. It introduced the concept of a European CI (ECI) where there is a cross-border common interest versus a national CI (NCI). In 2006, a proposal [38] was brought out for a directive and funding for the period 2007–2013. As is the case in the US, it stated with respect to the threats that terrorism would have priority but that the approach would be "all hazards". It further specified the principles to be followed in EU membership connection: subsidiarity, complementarity, confidentiality, stakeholder cooperation, and proportionality; while it foresaw a sector-by-sector approach.

In a press conference release [39] that same day in December 2006, 11 sectors were named:

1. energy
2. nuclear industry
3. information, communication technologies (ICT)
4. water
5. food
6. health
7. financial
8. transport
9. chemical industry
10. space
11. research facilities

It formulated what an ECI should be, what obligations the EC could impose on owners and operators in view of CIP, what kind of costs these should incur, and what the improvement should be.

The EC also redirected its Joint Research Centre (JRC) with its main location in Ispra, Italy, to support policies with respect to security in conjunction with safety.

This resulted in activities such as the coordination of the knowledge and research facilities sharing European reference network for critical infrastructure protection (ERNCIP), resilience analysis, and research for protection of communications and navigation with its space components, and physical protection of buildings.

In its preamble, the 2008 Council Directive [40] summarizes the evolution of EPCIP and how it is built on the premise that the Member States have the ultimate responsibility for the protection of the CIP within their borders, and that it is just the task of the EU in the common interest to identify transborder ECI. An ECI is an infrastructure that when severely damaged or destroyed will have an effect in the Member State (MS) where it occurs and will also have an impact on another MS. Effects will be possible fatalities and injuries, significant economic effects and loss of public confidence, physical suffering, or societal disruption. Identification of risks, threats, and vulnerabilities of ECIs is essential. This shall be performed by MSs and shared in a generic sense with the Commission. Rules of maintaining confidentiality both nationally and by the EU shall be followed.

Initiative in designating what an ECI is belongs to the MS where the CI is located, and this MS will start bilateral discussions with potentially affected other MSs in which the Commission may participate. In the case when an infrastructure is not designated as such, and another MS suspects it will be affected when this infrastructure is damaged, it can inform the Commission, which will attempt to initiate discussion.

For each ECI an Operator Security Plan will be drawn up according to a procedure described in an Annex to the Directive. Contact between the operator/owner of an ECI and the authority responsible in the MS shall be via a Security Liaison Officer. Within 1 year for each ECI, a threat assessment will be made and each 2 years risks, threats, and vulnerabilities in an ECI sector will be reported to the Commission. The sectors in which ECIs can be found are mentioned in Annex 1 of the Directive and are reproduced here in Table 2.3.

Tab. 2.3: ECI sectors mentioned in the 2008 EU CIP Directive [40]

Sector	Sub-sector		
Energy	1	Electricity	Infrastructures and facilities for generation and transmission of electricity in respect of supply electricity
	2	Oil	Oil production, refining, treatment, storage and transmission by pipelines
	3	Gas	Gas production, refining, treatment, storage and transmission by pipelines LNG terminals
Transport	4	Road transport	
	5	Rail transport	
	6	Air transport	
	7	Inland waterways transport	
	8	Ocean and short- sea shipping and ports	

In conclusion, the EU has established a framework for CIP and takes a coordinating role to protect against potential incidents with cross-border effects, both with respect to designation and exchange of information and state of the ECI. It is also coordinates and funds research, e. g., in 2016 within the Horizon 2020 EU Framework program research proposals were invited for CIP-01–2016-2017: prevention, detection, response, and mitigation of the combination of physical and cyber threats to the critical infrastructure of Europe. However, the majority of actual security measures is to be taken on a national basis.

The 2008 Directive [40] has been transposed in national law over the years following the entry into force in early 2009. In 2012, a review of the Directive was reported [41], showing that although there have been good examples, ECI identification could have been much better. As a side benefit the Directive increased the CIP importance awareness. Lazari [8] provides an overview of the titles of the national laws adopted in the various EU countries, commented on the implementation of the Directive, the fuzziness about its effectiveness, problems arising, and commented on a possible future role of the Commission. Activity on a revision of the directive commenced. The fundamental question has been asked again as to what sectors should be included. In 2011, the MS gave their opinion that it should be those sectors in which potential transnational and over countries cascading damaging events of CI can occur. This should include security of ICT, securing the dependency on space assets, e. g., navigation, securing infrastructure for financial transactions, food supply in view of transport and energy, and in the health sector protection against pandemics, securing pharmaceutical supplies, and other dependencies. In 2013, a Commission Working Document [42] appeared with a new approach to EPCIP proposing that the Commission should give support with respect to prevention (by risk assessment and management), to preparedness strengthening (contingency planning, stress tests, awareness raising, training, joint courses, exercises, and staff exchange), and response to weak signals. However, the limited mandate of the Commission does not give much leeway and given national political developments, one can question whether this leeway will become any wider.

Internet hacking activity can also be applied to disturb process control and interfere with programmable logic systems, including safety instrumented systems. Although of much wider interest than the limited scope of this chapter concerned with protection of process industry, it is good to sketch the main stream of policy developments with respect to ICT Cyber Security. It will also help to understand features in national regulation to be treated in the next chapter. ICT Cyber Security has had a special position in the EU since the foundation of the European Union Agency for Network and Information Security (ENISA) in 2004. Protection against cyber-attacks is a global issue. The Commission's communication in 2009 about Critical Information Infrastructure Protection (CIIP) [43] also contained, following initiatives in some MS, an announcement of a European Public-Private Partnership for Resilience (E3PR). This will make ENISA work together with national organizations and private telecom providers. One of the recommendations of the Centre for European Policy Studies (CEPS) in Brussels on CIP [44] in 2010 was also to build PPPs for trusted information sharing. By the way, other recommendations of CEPS, on first sight based on quite rational grounds, were to strive for a holistic and more centralized European approach of CI(I)P, which as we have seen above is in contrast with actual development.

Similarly to in the US after also having made a proposal, the EU issued early on a directive on port security. This became Directive/2005/65/EC [45].

2.4 Some European national solutions

Lazari [8] presented a table with the information on the implementation of the Directive into national law of the various MS (with exception of Estonia, Finland, and Ireland), which is here reproduced as Table 2.4. It will not be practical to describe solutions of all 27 EU MS in detail. Instead, two will be selected: Belgium and The Netherlands, which are two neighboring countries that have ECIs in common and have quite different approaches to the follow up of the 2008 Directive.

Belgium issued a special law on securing critical infrastructures [46] in 2011, following fairly closely the Directive. The law details the national organizational structure, designation of ECIs and NCIs, and relationships and points of contact with CI owners/operators. It prescribes the period in which a risk analysis identifying threat scenarios, a vulnerability analysis, and security plans are to be made and measures implemented. It regulates exchange and use of information, and competence of justice and police. The sectors of critical infrastructure mentioned in the law are electricity (power plants and transmission – nuclear power is treated separately; oil and gas – production, refining, storage, and pipeline transmission), transportation (road, rail, air, ship), the financial sector, and electronic communication. The process industry is only partially mentioned under the heading refining of oil and gas and it does not integrally include the chemical industry.

In 2006, The Netherlands founded a Contact group on Critical Infrastructure (SOVI = "Strategisch Overleg Vitale Infrastructuur") [47] in which several ministries, the industry association VNO-NCW, and each critical infrastructure sector are represented. This was followed in 2008 by the founding of the National Advice Centre Critical Infrastructure (NAVI = "Nationaal Adviescentrum Vitale Infrastructuur"), which was dissolved in 2010, while experts involved in NAVI founded a cooperative National Security Advisory Centre (NSAC) [48] for organizational and technical advice and risk assessment for government and industry. Expertise includes process industry security. An amendment to the Harbor Security law ("Havenbeveiligingswet", [49].) was announced in 2007.

Tab. 2.4: Implementation of EU Directive 2008/114/EC in National Law of MS [8][1]

Member State	Implementation measure
Austria	Amendment made to the national framework through specific administrative measures entered into force on January
Belgium	"Wet betreffende de beveiliging en de bescherming van de kritieke infrastructuren"; "Loi relative à la sécurité et la protection des infrastructures critiques"— entered into force on July 15, 2011
Bulgaria	Decree n. 18 "identifying and designating European critical infrastructures and the measures for their protection" entered into force on February 1, 2011
Cyprus	Regulations on the "Identification and Designation of European Critical Infrastructures and the Assessment of the Need to Improve their Protection", entered into force on January 20, 2011
Croatia	National law on Critical Infrastructures n. 56/2013
Czech Republic	Amendment of the ACT n. 240 on Crisis Management entered into force on June 28, 2000 through the Government Regulation No. 431/2010—amending Government Regulation No. 462/2000—and the Government Regulation No. 432/2010 "criteria for determining the elements of critical infrastructure"
Denmark	Promulgation of sector-specific Executive Orders: 1339/2007 (prevention of crimes against aviation security), 7/2011 (road-transport sector), 11/2011 (the identification and designation of European critical infrastructure in the energy sector), 1726/2010 (port security), 1461/2010 (railway sector), 6/2006 (ship domestic services)
France	Decree and the General Inter-ministerial Instruction N. 6600 SGDN/PSE/ PPS of September 26, 2008
Germany	National Laws revising the energy industry regulation (entered into force on August 4, 2011) and the protection of transmission systems (entered into force on January 10, 2012)
Greece	Adaptation of Greek legislation to the Directive 2008/114/EC through the Presidential Decree N. 39 entered into force on
Hungary	Resolution No. 1249/2010 of the Government of the Republic of Hungary on European Critical Infrastructures and the assessment of the need to improve their protection
Italy	Legislative Decree n. 61 entered into force on May 4, 2011
Latvia	Regulations N. 496 of the Cabinet of Ministers "Procedures for the Identification of Critical Infrastructures and European Critical Infrastructures"
Lithuania	Resolution N. 943 entered into force on August 24, 2011 and spector-specific Executive Orders
Luxembourg	"Règlement grand-ducal portant application de la directive 2008/114/CE du Conseil du 8 décembre 2008 concernant le recensement et la désignation des infrastructures critiques européennes ainsi que l'évaluation de la né cessité d'améliorer leur protection" entered into force on March 12, 2012

1 Estonia, Finland, and Ireland are missing from this table. Non-appearance does not mean that these countries are not active in CIP, on the contrary, but in their legislation the Directive is not easily traceable.

Tab. 2.4: (continued) Implementation of EU Directive 2008/114/EC in National Law of MS [8]

Member State	Implementation measure
Malta	Regulation N. 434 on "Critical Infrastructures and European Critical Infrastructures (Identification, Designation and Protection)" entered into force on November 8, 2011
The Netherlands	Amendment to the National CIP framework through the publication of the implementation program and requirement on the Official Gazette on the 23rd of December 2010
Poland	Act of October 29, 2010 on Crisis Management, Ordnances of the Council of Ministers of April 30, 2010 on the "national programme for Critical Infrastructure Protection" and "plans for the protection of critical infrastructures"
Portugal	Decree-Law N. 62 "procedures for the identification and protection of critical infrastructure for health, safety and economic, social well-being, energy and transport and transposing the Directive 2008/114/EC" entered into force on May 9, 2011
Romania	Emergency Ordinance on the identification, designation and protection of critical infrastructures, entered into force on November 16, 2010 and Government Decision on the composition, powers and organization of the Inter-institutional Working Group on Critical Infrastructure Protection entered into force on November 12, 2010
Slovakia	Act N. 45 on "Critical Infrastructures" entered into force on March 1, 2011
Slovenia	Decree n. 1799 on "European Critical Infrastructure" entered into force on May 12, 2011
Spain	Law N. 8/2011 for the Protection of Critical Infrastructure (entered into force on April 30, 2011) and Royal Decree N. 704/2011 "Regulation for the Protection of Critical Infrastructure" entered into force on May 22, 2011
Sweden	Ordinance N. 611-2009 amending the Ordinance 1002-08 "Swedish Civil Contingencies", Ordinance N. 513-2012 amending the Ordinance 1119-2007 "instruction to Swedish enterprises in the energy sector", Ordinance N. 793-2012 amending the Ordinance 185-2010 "instruction for the transport administration", Ordinance N. 512-2012 amending the Ordinance 1153-2007 "instruction for the Swedish Energy Agency"
UK	– Administrative Arrangement for amending the CPNI procedures in view to including those related to the assessment of the identification and designation of ECIs. – Gibraltar: amendment to the Civil Contingencies Act of 2007 (Gibraltar Gazette No. 3849 of May 12, 2011)

So far, we have treated CI in isolation and more specifically the physical protection aspect of it, and that in this chapter in relation to the potential hazards by the presence of process industry. We now come to a relatively new aspect of government policy that will hold also in the case of other countries but certainly in The Netherlands. Safety and security measures run parallel with respect to prevention of and protection from undesirable outcomes to people and the environment. People's risk level acceptability has decreased over the years, and the accountability of governments to provide safety and security has become more demanding. At the same time, due to lack of space for settling and hence smaller free distances, more intense use of traffic routes, increased industrial activities, intensified storms and enhanced chance of flooding resulting from climate change, higher terrorist threat, et cetera, both risk potential and exposure has significantly increased. Therefore, stimulated by international bodies such as UN and OECD, more emphasis is dedicated to disaster preparedness and management with overarching resilience building.

Thus, based on a proposal to parliament in 2006 and following a British example, in 2007 the Dutch government approved the National Strategy on Safety and Security ("Nationale Strategie Veiligheid", [50]). Critical infrastructure protection is just an element in this, also because of linguistic details as the Dutch noun "veiligheid," which encompasses both safety and security, while critical infrastructure is called vital infrastructure ("vitale infrastructuur") in Dutch. This designation may have an effect of lowering the visibility of CIP and the critical importance of resilience to achieve acceptable security. Working out the strategy led to a major activity in the years following of risk assessment of potential major disasters in the country resulting in a risk profile presented as a risk matrix per region and risk management to determine an optimal distribution of funding in preventive and protective measures, and organizational preparedness. A manual for determining the risk profile was published in 2009 [51], while the newest version is that of 2016 [52]. In contrast to, e. g., Belgium, this profile encompasses the process industry as a whole as it explicitly mentions "chemical disasters". The division into so-called safety regions ("veiligheidsregio's") has been in view of effective emergency response command and control. All this was established by a 2010 law (Wet Veiligheidsregio's, [53]). It can, therefore, be understood that the Directive 2008/114/EC [7] was not implemented by a separate law, but by an implementation action plan embedded in existing regulation. The plan was communicated in the Official Gazette ("Staatscourant", [54]) in 2010.

2.5 Conclusions

In the US, a broad and relatively intense effort has been displayed to improve the protection of the process industries. The legal structure has been built and the compliance effort is underway. It takes time and energy, but due to the online set p and the training of people involved, a high degree of efficiency is expected to be achieved. As inspection is part of the effort, in due time an overview of the state of affairs will exist on the federal level.

Europe followed the pattern with a few years' delay. As an EU activity it was able to initiate a reference framework and actual activity on cross border infrastructure, but the brunt of the effort was within the MS. EU directives have been absorbed by national law. The execution of the directives differs greatly between countries. Some have a specific law on security, others integrate security within existing regulation, aiming to be prepared for various disastrous events. An overview on the central level of threats and risks will exist only for the critical infrastructure designated as ECI. In Table 2.5 the mentioned differences between the US and European approaches are given in chronological order.

Although Europe is usually seen as a unity by people from outside the Union, this may not be the case inside it. A weakness in security coming to light by an event somewhere in an EU MS might be explained as a weakness in the whole EU. One can question whether a more centralized methodical approach with respect to the presence of hazardous materials, as developed in the US, would not increase overall effectiveness.

Tab. 2.5: Summary of similarities and differences in US and EU legislation in chronological order with respect to critical infrastructure protection with emphasis on process/chemical plants

Period	U.S.	E.U.
< 2001	Early recognition by President Clinton in 1998 that critical infrastructure protection deserves attention due to growing complexity, interdependence, and vulnerability to various kinds of threats. Critical infrastructure was defined as those physical and cyberbased systems essential to the minimum operations of the economy and government. A number of sectors were distinguished and an organizational structure founded in order to realize coordination and a warning communication network.	Before the 2001 WTC attack EU legislation with respect to safety had been fully developed, e.g., Seveso Directives, but no common regulation pertained to protection of critical assets. In general, security was kept by Member States at national level.
2002–2006	Immediately after the terrorist attack on the WTC buildings, under President Bush federal legislation was issued to found in 2002 the Federal Department of Homeland Security reshuffling ministerial structure and responsibilities to more effectively counter threats. In 2006 DHS issued its first National Infrastructure Protection Plan.	In 2004 the European Commission on request of the Council started working on a European Programme of Critical Infrastructure Protection (EPCIP). This was presented in 2006 with 11 sectors and a proposal for a directive. The proposal content character was much like that of the Clinton Presidential Policy Directive.

Tab. 2.5: (continued) Summary of similarities and differences in US and EU legislation in chronological order with respect to critical infrastructure protection with emphasis on process/chemical plants

Period	U.S.	E.U.
2007–2009	First actions focused on the transportation security of hazardous materials in the various modes of transport: maritime, road, rail, pipeline, and air. Already before the general awareness of CriticaL Infrastructure Protection (CIP), security of nuclear materials had been with the Department of Energy. To obtain the desired communication and coordination between departments Department of Homeland Security obtained the lead. Several Memorandums of Understanding were needed to obtain effective working procedures.	Despite the the proposed Directive, after all, the Member States wanted to retain their sovereignty in securing their own territory. *The EPCIP became limited to transnational hazards,* i.e., disruption of border crossing energy transmission or hazard effects from plants in the border area. The European Commission still has a role in stimulating and enabling coordination and communication. It can act as a higher authority in case a Member state does not fullfil its obligation in the framework of EPCIP, it also inspects and advises, and it funds R&D.
2010–2012	Recognizing the threat of terrorist act to process plant and chemical installations in particular to unleash damage threatening population, and the strategic supply of materials, and to steal hazardous chemicals, in 2006 the *Chemical Facility Anti-Terrorism Standards (CFATS)* Act was issued. This act obliges plant owners to have their installation classified with respect to risk and vulnerability in four tiers, and to have minimum protection measures installed, became operational in 2009 by the issue of a guidance document. Further instructions followed in 2011.	Besides implementing the 2008 EPCIP most Member States strengthened security measures on a national basis. *EPCIP does not mention process or chemical industry explicitly.* A number of countries, such as UK and The Netherlands embarked at about the same time on a program to identify all risks of disaster to the country in which critical infrastructure and terroristic threat were included. *Chemical risks* is part of it. This was to locate vulnerabilities and to plan, given the budget, preventive and protective measures including emergency response. To that end an overall risk matrix was constructed
2013–2016	Under President Obama the chosen directions were further extended and the number of critical infrastructure sectors increased to 16. The cyber threat became stronger during this period, e.g., by STUXNET type viruses. The execution of CFATS gradually took further shape and an acceleration of the execution in the tier 3 and 4 installations was realized.	In 2013 a new approach to EPCIP appeared proposing that the Commission should give support with respect to prevention (by risk assessment and management), to preparedness strengthening (contingency planning, stress tests, awareness raising, training, joint courses, exercises, and staff exchange), and response to weak signals. In the same year the Critical Infrastructure Warning Information Network (CIWIN) became operational. *No change was made with respect to chemical plants.*

Acknowledgment

The suggestions made with regard to the draft of this chapter by David A. Moore, PE, CSP of AcuTech consultants are highly appreciated.

References

References accessed July 2016

[1] Presidential Decision Directive/NSC-63. The White House Washington. May 22, 1998. http://fas.org/irp/offdocs/pdd/pdd-63.htm.

[2] Homeland Security Presidential Directive-1. October 29, 2001. https://fas.org/irp/offdocs/nspd/hspd-1.pdf.

[3] Critical Infrastructures Protection Act of 2001. 42 U. S. Code § 5195c – Critical Infrastructures Protection. https://www.law.cornell.edu/uscode/text/42/5195c.

[4] Homeland Security Act of 2002. Public Law 107–296—Nov 25, 2002. 116 Stat 2135. https://www.dhs.gov/xlibrary/assets/hr_5005_enr.pdf .

[5] Commission of the European Communities, Green Paper on A European Programme for Critical Infrastructure Protection (presented by the Commission). Brussels. Nov 11, 2005. COM(2005) 576 final. https://marcusviniciusreis.files.wordpress.com/2010/06/european-programe-to-protect-ci.pdf.

[6] Communication from the Commission of December 12, 2006 on a European Programme for Critical Infrastructure Protection [COM(2006) 786 final – Official Journal C 126 of 7.6.2007]. http://eur-lex.europa.eu/legal-content/EN/TXT/?uri=celex:52006DC0786.

[7] Council Directive 2008/114/EC of December 8, 2008 on the Identification and Designation of European Critical Infrastructures and the Assessment of the Need to Improve Their Protection. Official Journal of the European Union. December 23, 2008. L 345/75. http://geur-lex.europa.eu/LexUriServ/LexUriServ.do?uri=OJ:L:2008:345:0075:0082:EN:PDF.

[8] Lazari A. European critical infrastructure protection. Heidelberg: Springer (eBook).

[9] Bucci S, Inserra D. The Heritage Foundation. Issue Brief, No. 4072 October 23, 2013. http://thf_media.s3.amazonaws.com/2013/pdf/IB4072.pdf

[10] Department of Homeland Security. National Infrastructure Protection Plan – NIPP 2013. Partnering for Critical Infrastructure Security and Resilience. https://www.dhs.gov/publication/nipp-2013-partnering-critical-infrastructure-security-and-resilience.

[11] Title 6 – Domestic Security, Chapter 1 – Department of Homeland Security. Office of the Secretary Part 27 – Chemical Facility Anti-Terrorism Standards. April, 2007. https://www.gpo.gov/fdsys/pkg/CFR-2007-title6-vol1/pdf/CFR-2007-title6-vol1.pdf.

[12] Presidential Policy Directive/PPD-21, Critical Infrastructure Security and Resilience. https://www.whitehouse.gov/the-press-office/2013/02/12/presidential-policy-directive-critical-infrastructure-security-and-resil.

[13] The CIP Report. Center for Infrastructure Protection and Homeland Security. June 2015. http://cip.gmu.edu/wp-content/uploads/2013/06/155_The-CIP-Report-June-2015_EnergySector.pdf

[14] Implementation of Department of Energy Oversight Policy, Order 226.1B. Department of Energy, Washington, DC. April 4, 2011. https://www.directives.doe.gov/directives-documents/200-series/0226.1-BOrder-b.

[15] Independent Oversight Program, Order 227.1A. Department of Energy. Washington, DC. December 21, 2015. https://www.directives.doe.gov/directives-documents/200-series/0227.1-BOrder-A.

[16] Office of Enterprise Assessments, Independent Oversight Program, Appraisal Process Protocols. US Department of Energy. December 2015. http://energy.gov/ea/downloads/appraisal-process-protocols-independent-oversight-december-2015.
[17] Office of Enterprise Assessments, FY 2015 Independent Oversight Activities Overview, Report to Congress, October 2015. United States Department of Energy. Washington, DC 20585. http://energy.gov/sites/prod/files/2016/01/f28/2015 %20Annual%20Report%20to%20Congress%20 %28Final%29.pdf.
[18] Statement of Patricia Hoffman Assistant Secretary for Electricity Delivery and Energy Reliability U. S. Department of Energy Before the United States House of Representatives. Appropriations Subcommittee on Energy and Water Development. March 17, 2015. http://energy.gov/sites/prod/files/2015/07/f25/FY2016Budget-HEWD-testimony-3–17-15-OE-FINAL.pdf.
[19] Title XXI – Chemical Facility Anti–Terrorism Standards, Public Law 113–254. December 18, 2014. 128 STAT. 2898. https://www.congress.gov/113/plaws/publ254/PLAW-113publ254.pdf.
[20] Presidential Documents, Executive Order 13650 of August 1, 2013. Improving Chemical Facility Safety and Security, Federal Register /Vol. 78, No. 152 /Wednesday, August 7, 2013, 48029.
[21] U. S. Chemical Safety and Hazard Investigation Board. Investigation Report (final) West Fertilizer Company Fire and Explosion, Texas, April 17, 2013. Report 2013–02-I-TX. January 2016. http://www.csb.gov/.
[22] U. S. Department of Homeland Security. 6 CFR Part 27. Chemical Facility Anti-Terrorism Standards. Federal Register/Vol 72, No 67/Monday, April 9, 2007/Rules and Regulations, 17688. http://www.ecfr.gov/cgi-bin/text-idx?tpl=/ecfrbrowse/Title06/6cfr27_main_02.tpl.
[23] U. S. Department of Homeland Security. Risk-Based Performance Standards Guidance, Chemical Facility Anti-Terrorism Standards. May 2009. https://www.dhs.gov/chemical-facility-anti-terrorism-standards.
[24] U. S. Department of Homeland Security. CSAT Security Vulnerability Assessment, Questions. June 2008. Version 1.0, OMB PRA # 1670–0007. https://www.dhs.gov/chemical-security-assessment-tool.
[25] U. S. Department of Homeland Security. CSAT Security Vulnerability Assessment Application, Instructions. January 3, 2011. Version 2.1. https://www.dhs.gov/chemical-security-assessment-tool.
[26] US Department of Homeland Security. CSAT Site Security Plan, Questions. June 2011. Version 2. https://www.dhs.gov/chemical-security-assessment-tool.
[27] U. S. Department of Homeland Security. CSAT Site Security Plan, Instructions. May 2009. Version 1.0. https://www.dhs.gov/chemical-security-assessment-tool.
[28] U. S. Department of Homeland Security. Chemical Sector Security Awareness Guide. A Guide for Owners, Operators, and Chemical Supply-Chain Professionals. September 2012. https://www.dhs.gov/sites/default/files/publications/DHS-Chemical-Sector-Security-Guide-Sept-2012–508.pdf.
[29] Shea DA. Implementation of chemical facility anti-terrorism standards (CFATS): issues for Congress. Congressional Research Service. April 2014. 7–5700. www.crs.gov R43346.
[30] U. S. Department of Homeland Security. DHS Guidance for the Expedited Approval Program. https://www.dhs.gov/sites/default/files/publications/DHS-EAP-Guidance-Document-05–15-508.pdf.
[31] Maritime Transportation Security Act of 2002. Public Law 107–295. November 25, 2002. 116 Stat. 2064. https://www.gpo.gov/fdsys/pkg/PLAW-107publ295/pdf/PLAW-107publ295.pdf.
[32] 33 CFR Chapter 1. Sub-Chapter H – Maritime Security. http://www.ecfr.gov/cgi-bin/text-idx?tpl=/ecfrbrowse/Title.
[33] Homeland Security Presidential Directive No. 7. December 17, 2003. Critical Infrastructure Identification, Prioritization and Protection (HSPD-7). https://www.dhs.gov/homeland-security-presidential-directive-7.

[34] Annex to the Memorandum of Understanding between the Department of Homeland Security, and the Department of Transportation concerning Transportation Security Administration and Pipeline and Hazardous Materials Safety Administration Cooperation on Pipeline and Hazardous Materials Transportation Security. http://www.phmsa.dot.gov/staticfiles/PHMSA/DownloadableFiles/Annex%20to%20MOU%20between%20TSA-PHMSA.PDF.
[35] Memorandum of Understanding among the Department of Homeland Security, the Department of Transportation, the U. S. Nuclear Regulatory Commission Concerning Cooperation on Radioactive Materials Transportation Security. Last signature 2015. http://pbadupws.nrc.gov/docs/ML1505/ML15057A336.pdf.
[36] 49 CFR Transportation. http://www.ecfr.gov/cgi-bin/text-idx?tpl=/ecfrbrowse/Title49/49tab_02.tpl.
[37] Commission of the European Communities. Green Paper on a European Programme Critical Infrastructure Protection. Brussels, November 17, 2005. COM(2005) 576 final. https://marcusviniciusreis.files.wordpress.com/2010/06/european-programe-to-protect-ci.pdf.
[38] Commission of the European Communities. Communication from the Commission on a European Programme Critical Infrastructure Protection. COM(2006) 786 final. December 12, 2006. http://eur-lex.europa.eu/LexUriServ/LexUriServ.do?uri=COM:2006:0786:FIN:EN:PDF.
[39] Press Conference Release. The European Programme for Critical Infrastructure Protection (EPCIP), MEMO/06/477. Brussels. December 12, 2006. http://europa.eu/rapid/press-release_MEMO-06–477_en.htm.
[40] Council Directive 2008/114/EC of 8 December 2008. On the Identification and Designation of European Critical Infrastructures and the Assessment of the Need to Improve Their Protection. December 23, 2008. EN Official Journal of the European Union. L 345/75. http://eur-lex.europa.eu/LexUriServ/LexUriServ.do?uri=OJ:L:2008:345:0075:0082:EN:PDF.
[41] European Commission. Commission Staff Working Document on the Review of the European Programme for Critical Infrastructure (EPCIP). Brussels. June 22, 2012. SWD(2012) 190 final. http://ec.europa.eu/dgs/home-affairs/what-we-do/policies/crisis-and-terrorism/critical-infrastructure/index_en.htm.
[42] European Commission: Commission Staff Working Document on a New Approach to the European Programme for Critical Infrastructure Protection. Making European Critical Infrastructures More Secure. Brussels. August 28, 2013. SWD(2013) 318 final,. http://ec.europa.eu/dgs/home-affairs/what-we-do/policies/crisis-and-terrorism/critical-infrastructure/index_en.htm.
[43] Communication from the Commission to the European Parliament, the Council, the European Economic and Social Committee and the Committee of the Regions on Critical Information Infrastructure Protection. Protecting Europe from Large Scale Cyber-Attacks and Disruptions: Enhancing Preparedness, Security and Resilience. Brussels. March 30, 2009. COM(2009) 149 final. http://eur-lex.europa.eu/LexUriServ/LexUriServ.do?uri=COM:2009:0149:FIN:EN:PDF.
[44] Hämmerli B, Renda A. Protecting Critical Infrastructure in the EU. CEPS Task Force Report. Centre for European Policy Studies, Brussels. © CEPS 2010. https://www.ceps.eu/publications/protecting-critical-infrastructure-eu.
[45] Directive 2005/65/EC of the European Parliament and of the Council, of 26 October 2005 on Enhancing Port Security. L 310/28 EN Official Journal of the European Union. November 25, 2005. http://eur-lex.europa.eu/legal-content/EN/TXT/PDF/?uri=CELEX:32005L0065&rid=1.
[46] Loi relative à la sécurité et la protection des infrastructures critiques – Wet betreffende de beveiliging en de bescherming van de kritieke infrastructuren. July 1, 2011. English translation http://www.microsofttranslator.com/bv.aspx?ref=SERP&br=ro&mkt=nl-NL&dl=en&lp=NL_EN&a=http%3a%2f%2fwww.ejustice.just.fgov.be%2fcgi_loi%2floi_a1.pl%3flanguage%3dnl%26la%3dN%26cn%3d2011070108 %26table_name%3dwet%26 %26caller%3dlist%26fromtab%3dwet%26tri%3ddd%2bAS%2bRANK.

[47] Besluit instelling van het Strategisch Overleg Vitale Infrastructuur (Instellingsbesluit SOVI), geldend van 28–04-2006 t/m heden. http://wetten.overheid.nl/BWBR0019781/2006–04-28.
[48] National Security Advisory Centre (NL: Adviescentrum BVI). The Hague, The Netherlands. http://www.nsac.eu/default.htm.
[49] Wijzigingswet Havenbeveiligingswet (implementatie richtlijn nr. 2005/65/EG betreffende verhogen veiligheid van havens), geldend van 01–10-2010 t/m heden. http://wetten.overheid.nl/BWBR0016991/2010–10-01.
[50] Tweede Kamer der Staten-Generaal, Vergaderjaar 2006–2007. 30821 Nationale Veiligheid. Brief van de Minister van Binnenlandse Zaken en Koninkrijkrelaties aan de Voorzitter van de Tweede Kamer der Staten-Generaal, Den Haag. October 2, 2006. KST101379, 0607tkkst30821,-1 ISSN 0921 – 737. Sdu Uitgevers, 's-Gravenhage 2006. https://zoek.officielebekendmakingen.nl/kst-30821–3.pdf.
[51] Handreiking Regionaal Risicoprofiel, Politie, NVBR, GHOR, Coördinerend Gemeentesecretarissen. 5 November, 2009 (EN: Guideline Regional Risk Profile – Police, Fire Brigade Organization, Regional Medical Organization, Coordinating Municipal Secretaries). http://www.regionaalrisicoprofiel.nl/algemene_onderdelen/downloads/handreiking/.
[52] Nationaal Veiligheidsprofiel 2016. Een All Hazard overzicht van potentiële rampen en dreigingen die onze samenleving kunnen ontwrichten. Analistennetwerk Nationale Veiligheid, © RIVM 2016 (National Safety and Security Profile 2016. An All Hazard Overview of Potential Disasters and Threats Potentially Able to Disrupt Society. Analyst Network National Safety and Security). https://zoek.officielebekendmakingen.nl/blg-793151.pdf.
[53] Wet van 11 februari 2010. Houdende bepalingen over de brandweerzorg, de rampenbestrijding, de crisisbeheersing en de geneeskundige hulpverlening (Wet veiligheidsregio's), geldend van 01–01-2016 t/m heden. http://wetten.overheid.nl/BWBR0027466/2016–01-01.
[54] Mededeling inzake de implementatie van richtlijn 2008/114/EG. Staatscourant Nr. 20996. December 24, 2010. https://zoek.officielebekendmakingen.nl/stcrt-2010–20996.html?zoekcriteria=%3fzkt%3dUitgebreid%26pst%3dStaatscourant%26dpr%3dAnderePeriode%26spd%3d20101224 %26epd%3d20160717 %26nrp%3d%252020996 %26sdt%3dDatumPublicatie%26planId%3d%26pnr%3d1 %26rpp%3d10&resultIndex=6&sorttype=1&sortorder=4.

Paul Baybutt

3 Security vulnerability analysis: protecting process plants from physical and cyber threats

Glossary of terms

Accident: An incident with adverse consequences that happens unexpectedly and unintentionally.
Adversary: An individual, group, or organization that decides to target assets of others.
Asset: An entity that has value to its owner(s) and/or attacker(s).
Attacker: See adversary.
Consequence: The outcome of an event or series of events.
Control: See countermeasure. Alternatively, the means to keep a process within operating limits.
Countermeasure: Safeguard or secureguard.
Criticality: The importance of a threat.
Harm: Adverse impacts to an asset.
Hazardous event: An accidental occurrence that can cause harm.
Intent: The combination of the objective of an attack and the means of achieving it using an asset.
Likelihood: The chance of something happening.
Malevent: An incident with adverse consequences resulting from a deliberate act performed with the intention of causing harm.
Mitigation: Action that reduces the consequence(s) of an event.
Prevention: Action that reduces the likelihood of occurrence of an event.
Protection layer: An independent mechanism that reduces risk by control, prevention or mitigation.
Risk: A measure of the consequences and likelihoods of an event or set of events. A fuller definition includes the expression of uncertainties.
Safeguard: A measure to protect against a hazardous event.
Safety: Freedom from the risk of hazardous events that is not tolerable.
Secureguard: A measure to protect against a malevent.
Security: Freedom from the risk of malevents that is not tolerable.
Target: A facility, place, object, or person that is the objective of an attack.
Threat: The possibility of hostile action towards an asset by an adversary with the intention of causing harm. Alternatively, attacker plus intent plus capability.
Threat event: The pairing of an adversary and their intent with a targeted asset.
Threat scenario: A specific sequence of events with an undesirable consequence resulting from the realization of a threat.
Tolerable risk: Level of risk that is accepted in a given context based on the current values of society.
Vulnerability: Flaws or weaknesses that can be exploited by an adversary to gain access to an asset.

https://doi.org/10.1515/9783110499087-003

Abbreviations

ACC	American Chemistry Council
CSVA	Cyber SVA
IT	Information technology
LOPA	Layers of protection analysis
PHA	Process hazard analysis
QRA	Quantitative risk analysis
ROPA	Rings of protection analysis
SVA	Security vulnerability analysis

3.1 Introduction

Many process plants contain hazardous materials that, if released, can adversely impact the health and safety of workers and the public, and damage the environment or property. Such releases can result from extraordinary events such as accidents, natural events, or deliberate acts (Figure 3.1). Accidents occur when people make errors or mistakes, or equipment fails. Natural events are phenomena such as lightning strikes and flooding, sometimes called external events. Deliberate acts, called *malevents* herein, are performed with the intention of causing harm and include terrorism, sabotage, vandalism, and theft. Harm may include the release or diversion of hazardous materials and process or product damage. Some facilities may not contain hazardous materials but may be part of the critical infrastructure for society, for example, pharmaceutical manufacturing. Deliberate acts that damage or disable such facilities also are of concern. Malevents are the security equivalent of an accident.

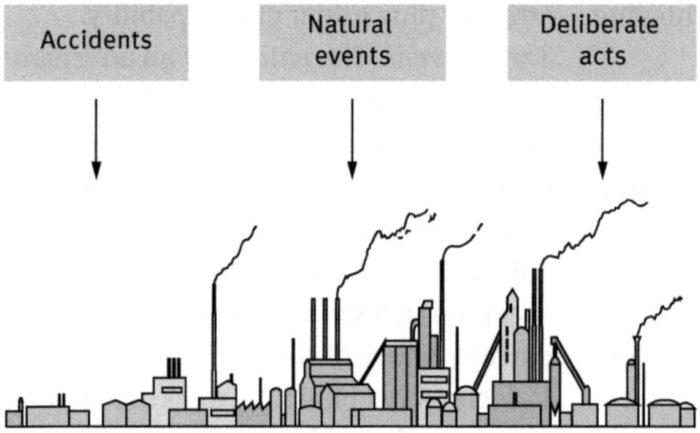

Fig. 3.1: Extraordinary events for a process plant.

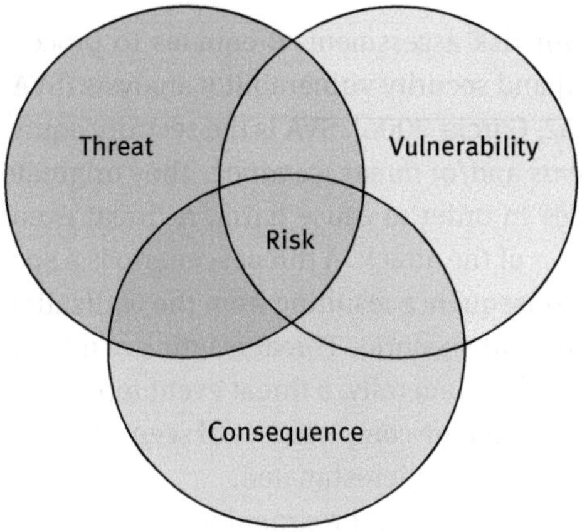

Fig. 3.2: View of security risk

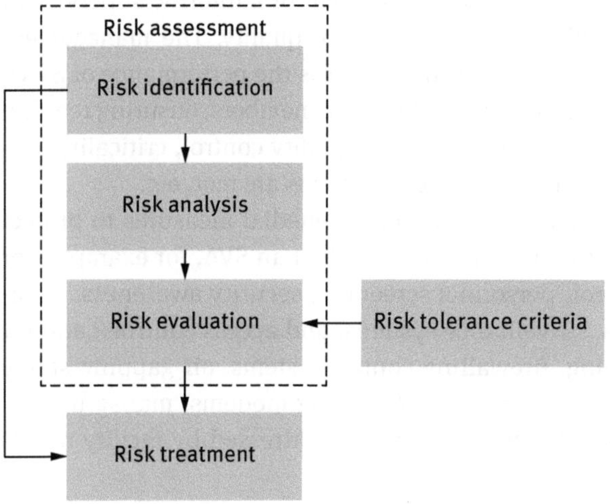

Fig. 3.3: Risk assessment

The risks of such deliberate acts are addressed using security risk assessment (SRA) to determine if existing security measures and process safeguards are adequate or need improvement [41]. Conceptually, security risk can be viewed as the intersection of events where threat, vulnerability, and consequences are present (Figure 3.2). Risk assessment consists of risk identification, risk analysis, and risk evaluation (Figure 3.3) [44, 45]. *Risk identification* involves the identification of risk sources, events, their causes, and potential consequences. *Risk analysis* is used to determine the level of risk. *Risk evaluation* is the process of comparing the results of risk analysis with risk criteria to determine whether the risk is tolerable. It assists in the decision about *risk treatment* to reduce risk, if needed. Note that risk identification may lead directly to risk treatment.

Risk identification is the starting point for risk assessment. It equates to process hazard analysis PHA in the safety domain [20] and security vulnerability analysis (SVA) in the security domain [2, 5, 6, 7, 9, 35, 38, 40, 52, Garcia 2006]. SVA is the security equivalent of PHA. It involves evaluating *threat events* and/or *threat scenarios*. They originate with hostile action to gain access to processes in order to cause harm. A threat event pairs an attacker and their intent with the object of the attack. A threat scenario is a specific sequence of events with an undesirable consequence resulting from the realization of a threat. It is the security equivalent of a hazard scenario. Threat events are a higher-order representation of detailed threat scenarios. Generally, a threat event represents a set of threat scenarios. Risk assessment depends on the completeness of scenario identification in SVA. If scenarios are missed, risks will be underestimated.

SVA should be part of an overall security management program for a facility [10], which companies should design, develop, implement, operate, maintain, and continuously improve. A key aspect of such a program is a management system that addresses the assignment of roles and responsibilities, the provision of authority, supervision, allocation of resources, accountability, and assurance of quality. The management system should include policies and procedures that address the performance of SVAs, including acceptable approaches, the qualification of team members, ensuring requirements for team composition are met, requirements for quality control, criticality estimation, ensuring recording and documentation requirements are met, etc.

Prior to performing SVAs, companies should take remedial measures to protect their facilities that are obvious without the need to conduct an SVA, for example, for physical security: inventory control, personnel screening, security awareness, information control, physical barriers, surveillance systems, and access controls; and for cyber security: personnel screening, firewalling control systems, air gapping safety instrumented systems, eliminating or controlling/securing modems, managing portable computer storage media, etc. Such issues can be addressed by facility audits before SVAs are performed.

3.2 Nature of threats

Threats represent the possibility of hostile action, which can take various forms. Process plants may be subject to physical or cyber attacks or a combination thereof.

3.2.1 Physical threats

Physical attacks target assets within a facility or process such as inventories of hazardous materials. They may be mounted on a facility by attackers trying to reach the assets by penetrating the facility or reaching the assets remotely from outside the facility perimeter. Physical attacks may result in:

- Release of hazardous materials
- Theft or diversion of materials
- Contamination of chemicals, materials, or products
- Damaging, destroying, or stealing assets
- Manipulating or disabling equipment, processes, plants, or other assets

3.2.2 Cyber threats

Process plants use computer systems to control manufacturing processes and safety systems operation, store information, manage value chain activities, etc. In modern plants, these computer systems often are connected to other networks driven by the need to communicate process information to business groups. This exposes the systems to access by more people and access through the Internet. All these computer systems and their support systems are subject to threats, including the following:

- Manipulation of process equipment such as pumps, valves, and motors to cause a hazardous material release, runaway reaction, diversion of materials, contamination or poisoning of products, etc.
- Misdirecting material transfers.
- Modification of set points for such process parameters as pressure, temperature, and level.
- Disabling or overriding alarms and trip settings.
- Disabling interlocks and safety instrumented systems.
- Disabling visual display units that are required for safe process operation.
- Disabling, damaging or destroying cyber assets to prevent their proper operation or cause a financial loss.
- Loss, theft, disclosure, damage, destruction, corruption, or prohibition of access to valuable data or information stored in cyber assets.

A cyber-attack may be mounted to obtain sensitive information to plan a future physical or cyber-attack.

Not all cyber events are malicious. They can also be caused by accident. People may make mistakes such as incorrectly entering data, using the wrong data, accessing the incorrect system, misprogramming systems, using conflicting software, etc. These accidental risks should be assessed as part of process hazard analysis.

3.3 Threat events and threat scenarios

A threat event is a higher level representation of a threat scenario. For example, "release of chlorine by a disgruntled employee" is a threat event, while "release of chlorine by a disgruntled employee opening manual valves" is a threat scenario.

Usually, there will be multiple threat scenarios that correspond to each threat event. Threat scenarios involve various elements (see Figure 3.4).

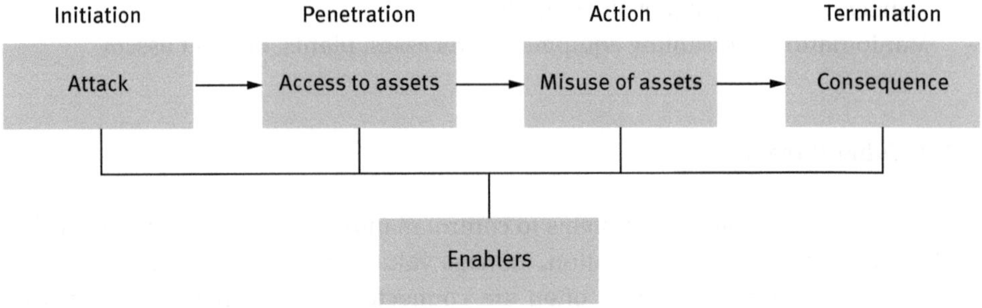

Fig. 3.4: Elements of a threat scenario.

Adversaries, attackers, or *assailants* may be individuals, groups, or organizations that conduct activities deliberately, or have the intention and capability to conduct activities, to attack assets. They may include insiders, for example, disgruntled employees, contractors, customers, vendors, or others who have some measure of unrestricted access to a facility; or outsiders, for example, terrorists, saboteurs, hostile foreign governments, criminals, hackers, or activists who do not. Collusion between insiders and outsiders may occur from financial motivation, ideological sympathy, or coercion. Contractors can cover many different categories, including electrical, mechanical, and other types of maintenance; repair service technicians; nondestructive testing technicians; painting contractors; food service workers; vending machine operators; landscaping contractors; janitorial staff; cleaning crews; guards, etc. Insiders know where assets are located, have access to them, and know what to do with them. They can cause serious damage.

Certain adversaries may be more likely to target a particular asset. For example, cyber terrorists may target computer control systems if they believe manipulation could result in plant shutdown or a release of hazardous material. Adversaries intent on stealing chemicals may target the value chain and competitors may be more interested in information technology (IT) systems.

An *attack* is hostile action taken by an adversary to obtain access to an asset and use it to cause harm. Typical attack objectives are to deny the use of the asset, damage or destroy it, or divert it to some other purpose. Objectives may include the release of hazardous materials; the theft of chemicals for later use as weapons, or other misuse; the contamination of chemicals or tampering with a product that may later harm people; the damage or disruption to a plant or process; and the disruption of everyday life.

Assets are entities that have value to someone. They may be tangible or intangible. Examples are people, materials, equipment, an activity or operation such as manufacturing, information, business reputation, the environment, and the community. Assets have value both to the owner and to adversaries, but for different reasons. They are of value to a company when they are needed to conduct operations. They are of value to an adversary when they can be used to inflict harm, either to their owners or others.

Key chemical assets in process plants may include inhalation poisons and asphyxiants, water soluble toxics, flammables, reactives, explosives, oxidizers and explosive precursors, corrosives, carcinogens, chemical weapons precursors, chemicals that act as blister or nerve agents, water supply contaminants, environmental contaminants, and chemicals important to the national, public, industry or company economic or other interest, including national security. Attention should be paid to raw material unloading and product loading stations and liquefied or pressurized toxic or flammable gas systems.

Key equipment in processes are vessels such as storage tanks, reactors, and fractionation columns, and piping that contains significant inventories of chemicals. Equipment items such as valves and pumps are subject to manipulation. Key buildings in process plants are control rooms and warehouses.

Key utility assets in process plants include electrical generating stations and substations; backup power generation systems; natural gas and other fuels; steam; nitrogen or other inert gases; instrument air supply; process heating and cooling systems; process, potable, and firewater supply systems; drainage and sewer systems; and heating, ventilation, and cooling systems.

Key computer assets in process plants include those used for manufacturing and process control, safety systems operation, utility operation, maintenance, communications, facility access, information management and storage, value chain activities, and enterprise and business management. They include distributed control systems (DCSs), programmable logic controllers (PLCs), supervisory control and data acquisition (SCADA) systems, mainframe computers, minicomputers, and personal computers (PCs). Both basic process control systems and safety instrumented systems must be considered. Cyber assets include hardware, software, data, and peopleware (the people who interact with them).

Locations that need to be protected for computers include computer rooms, server rooms, process control rooms and stations, utility control rooms and stations, motor control centers, rack rooms, and telecommunications rooms. Computer support systems such as utilities, for example, electric power and backup power, and fire protection should also be addressed. Business computer systems are typically located at corporate headquarters, individual plant sites, or other locations such as supplier, customer, service provider, or contractor facilities. Computer control systems are usually located at an individual plant site or at a remote location. Value chain computer systems may be at any of these locations or they may be mobile, for example, on trucks.

Assets may or may not be owned by a company, but any assets under the control of a company or that are integrated into a company's operations should be considered. This includes, for example, road tankers or rail cars on site making deliveries, adjacent facilities that are integrated into a company's operations or that could have a significant impact on the facility, and computer systems operated at vendor sites that contain sensitive company information or can be used to cause harm if connected to company networks.

Vulnerabilities are flaws or weaknesses that can be exploited by an adversary to successfully attack an asset. Examples of physical vulnerabilities are unlocked gates, lack of intrusion detection, and unrestricted access to control rooms. Examples of cyber vulnerabilities are dial-up modems for a control system, lack of encryption for value chain activities, and weak passwords for business systems.

Countermeasures, also called *controls,* provide protection against malevents. Generally, security protection tries to prevent physical and cyber access to facility assets while safety protection tries to prevent failure or misoperation of a process that could cause harm. In process safety, the term *safeguards* is usually intended to convey measures to protect against accidents. In process security, various security measures that do not necessarily assist in protecting against accidents are needed to protect against threats. These can be called *secureguards.* Countermeasures are secureguards that address the security of systems and safeguards that help ensure systems remain safe from attack. In process security management, safeguards and secureguards must be combined into a program to provide overall protection [8].

Consequences are the impacts of attacks. They may affect people, property, the environment, processes, products, companies, local communities, society, the nation, etc. SVA identifies these impacts. The impacts of attacks depend on the type of attack. These will likely vary for different systems. For example, physical attacks on hazardous material storage tanks will likely impact facility personnel and the public, while attacks to steal valuable chemicals may impact only the company. Similarly, cyber attacks on business computer systems will likely impact primarily the company, while attacks on computer control systems could impact the public if they are manipulated to cause hazardous material releases, as could attacks on the value chain if hazardous materials are diverted.

Recommendations are suggested actions that can be taken to reduce the risk of malevents to tolerable or acceptable levels. Decisions on the need for new or improved countermeasures are based on the nature of the threat, vulnerabilities present, existing protective measures, and the magnitude and type of consequences. SVA can include risk estimation to assist in decision making by assessing the likelihood and severity of threat scenarios. However, such risk estimation can be problematic and is discussed in a later section.

3.4 Overview of SVA

SVA usually addresses high-risk events with potentially catastrophic consequences, such as those that may arise as a result of terrorist attacks. Typically, these involve large-scale impacts that could affect a significant number of people, the public, the facility, the company, the environment, the economy, or the country's infrastructure (industrial sectors needed for the operation of the economy and government). However, SVA can also be used to address other plant security risks, such as the theft of valuable process information for financial gain.

An SVA for a facility endeavors to address the following questions:
- Will a facility be targeted?
- What assets may be targeted?
- How may assets be exploited?
- Who will attack?
- How will they attack?
- What protection is there against an attack?
- What will be the consequences?
- Is additional protection needed?

The overall objectives of SVA are to identify credible threats to a facility, identify vulnerabilities that exist, and provide information to facilitate decisions on any corrective actions that should be taken. SVA uses structured brainstorming by a team of qualified and experienced people, a technique that has a long history of success in the safety field. It has been noted that identifying scenarios for risk analysis is part science and part art [47]. SVA requires the application of creative thinking [27] to help ensure the completeness of threat and vulnerability identification and critical thinking [26, 51] to help ensure that the results are not subject to cognitive or motivational biases [24, 50]. The underlying model for the analysis is depicted in Figure 3.5.

SVA can be applied both to continuous and batch processes. The risks may vary according to the state of the process, for example, startup, normal operation, shutdown, etc., owing to changes in forms of process chemicals, process conditions, state of equipment, etc. This is particularly true for batch processes where the risks may vary according to the step in the batch. These variations should be addressed, either by considering each stage in the process life cycle individually, or by considering worst-case situations.

Facility assets and threats can be screened to determine specific types of attack to consider when identifying vulnerabilities. It is also possible for the SVA team to do so when identifying vulnerabilities.

Various SVA methods have been developed to identify and analyze threats and vulnerabilities of process plants to attacks. They share a number of points in common and they all address assets to be protected. They differ only in the approach taken.

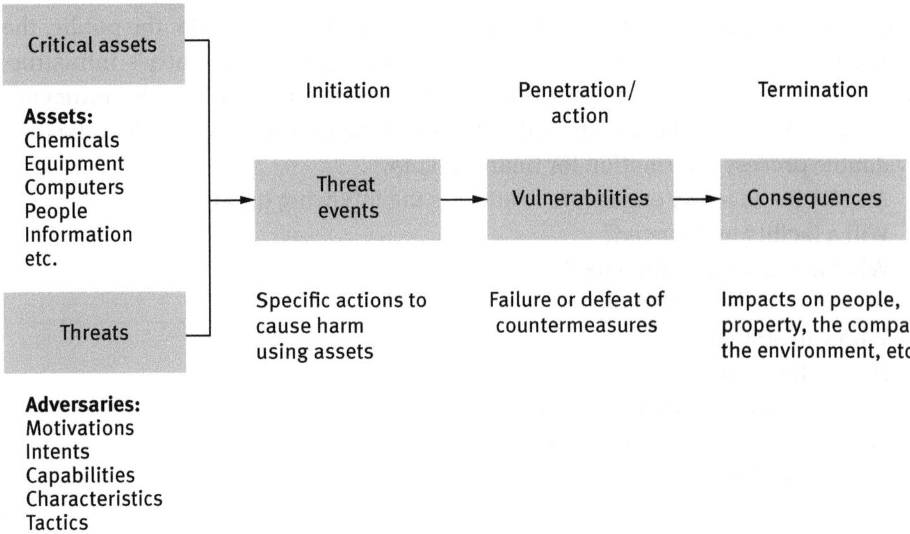

Fig. 3.5: SVA model

3.5 Types of SVA

Historically, two philosophically different SVA approaches were developed for physical security: asset based and scenario based [35]. The asset based approach originated with security professionals who focused efforts on protecting valuable assets. The scenario based approach originated with safety professionals who focused on protecting against accidents and the scenarios they involve. Both approaches consider how assets can be exploited by adversaries to cause harm. Asset based and scenario based approaches for cyber security have also been developed [12, 14].

A sneak path method has also been developed and applied to cyber security [15]. All three methods have been compared and contrasted [17]. These methods were evaluated by the Chemical Industry Data Exchange on behalf of the American Chemistry Council (ACC) for application to cyber security [37]. Asset based, scenario based, and sneak path SVA can be used to address physical and/or cyber security. All approaches look for vulnerabilities or weaknesses in the system that may allow successful attacks to occur. Existing countermeasures are identified, and the need for new or improved countermeasures is considered.

SVA methods are performance based and do not require the use of any specific risk remediation measures or countermeasures. SVA studies must be documented to allow review by peers and others. Often, SVA study results are recorded in the form of a spreadsheet, which offers the benefit of easy updating when needed. The format of the analyses is similar to PHA and, therefore, the methods offer the further benefit of familiarity to individuals who have participated in PHAs; a number of whom will likely also be members of SVA teams.

3.5.1 Asset based SVA

Asset based SVA pairs threats with assets to define threat events and focuses protective measures on assets. The method provides results quickly and identifies overall protective measures needed. Examples of physical and cyber asset based SVA worksheets are provided in Tables 3.1 and 3.2.

The level of detail in asset based SVA can be varied by breaking down assets into more detail. For example, individual PLCs could be listed in Table 3.2.

Tab. 3.1: Example of asset based physical SVA

ASSETS	ATTACKERS	INTENTS	CONSEQUENCES	RECOMMENDATIONS
Chlorine	Disgruntled employee	Release	Multiple fatalities on-site and off-site	Consider locking manual valves
				Consider installing an alarm for public notification of release
	Terrorist	Release	Multiple fatalities on-site and off-site	Consider installing CCTV surveillance
				Consider fencing tank farm and providing intrusion detection system
Ammonia	Disgruntled employee	Release	Multiple fatalities on-site	Consider locking manual valves

Tab. 3.2: Example of asset based cyber SVA

ASSETS	ATTACKERS	INTENTS	CONSEQUENCES	RECOMMENDATIONS
PLCs	Hacker	Operate equipment and cause a chemical release	Possible on-site fatalities	Consider use of biometric authentication
				Consider installing an intrusion detection system
		Disable computer system	Loss of production	No additional recommendations
Control room	Terrorist	Use control system to cause a chemical release	Possible on-site and off-site fatalities	Provide access controls
				Harden control room
Dial-in modem	Hacker	Operate equipment and cause a chemical release	Possible on-site fatalities	Consider eliminating modem
				Use secure modem
		Disable computer system	Loss of production	No additional recommendations
Server	Insider	Create problems for the company	Operational difficulties	Provide access controls

3.5.2 Scenario based SVA

Scenario based SVA develops threat events into more detailed threat scenarios for analysis and focuses protective measures on the threat scenarios. Countermeasures are addressed for each scenario. The method requires more time and effort than asset based SVA but provides more detailed recommendations for protective measures. Examples of physical and cyber scenario based SVA worksheets are provided in Tables 3.3 and 3.4. The level of detail usually is greater in scenario based SVA than in asset based SVA. However, the level of detail has been kept the same in the examples provided here to facilitate a comparison of the methods.

Tab. 3.3: Example of scenario based physical SVA

ASSETS	ATTACKERS	INTENTS	VULNERABILITIES	CONSEQUENCES	COUNTERMEASURES	RECOMMENDATIONS
Chlorine	Disgruntled employee	Release	Manual valves can be opened	Multiple fatalities on-site and off-site	Gas detectors	Consider locking manual valves
					Tank farm operator in area	Consider installing an alarm for public notification of release
					HAZMAT response team	
			Control system can be used to open valves	Multiple fatalities on-site and off-site	Access to control room is restricted	
			Safety system to prevent overfilling can be disabled	Multiple fatalities on-site and off-site	Set points can be changed only by lead operator	
	Terrorist	Release	Tank is close to boundary fence	Multiple fatalities on-site and off-site	Guard patrols	Consider installing CCTV surveillance
						Consider fencing tank farm and providing intrusion detection system

Tab. 3.4: Example of scenario based cyber SVA

ASSETS	ATTACKERS	INTENTS	VULNERABILITIES	CONSEQUENCES	COUNTERMEASURES	RECOMMENDATIONS
PLCs	Hacker	Operate equipment and cause a chemical release	No user authentication	Possible on-site fatalities	Network firewall	Consider use of bio-metric authentication
					Release detection and response	Consider installing an intrusion detection system
		Disable computer system		Loss of production	Network firewall	No additional recom-mendations
Control room	Terrorist	Use control system to cause a chemical release	No restrictions on access to control room	Possible on-site and off-site fatalities	Control room is cen-trally located	Provide access controls
						Harden control room
Dial-in modem	Hacker	Operate equipment and cause a chemical release	Weak password on modem	Possible on-site fatalities	Release detection and response	Consider eliminating modem
						Use secure modem
		Disable computer system		Loss of production	None identified	No additional recom-mendations

3.5.3 Sneak path SVA

Sneak path SVA is used to identify the paths or ways (vulnerabilities) in which attackers may access assets to cause harm. Paths may exist as design flaws and correspond to the latent conditions of conventional sneak path analysis. Other paths may require the breaching of existing barriers and are considered using barrier analysis in the sneak path analysis. The method provides a conceptually simple framework for conducting SVA. Examples of physical and cyber sneak-path SVA worksheets are provided in Tables 3.5 and 3.6.

Tab. 3.5: Example of sneak path physical SVA

ATTACKERS	TARGETS	PATHS	BARRIERS	EVENTS	CONSEQUENCES	RECOMMENDATIONS
Disgruntled employee	Chlorine storage	Manual valves	Tank farm operator in area	Chlorine release	Multiple fatalities on- site and off-site	Consider locking manual valves
		Control system can be used to open valves	Access to control room is restricted			Consider restricting access to control room
		Override of safety on filling system	Set points can be changed only by lead operator			Consider implementing an oversight policy
Terrorist	Chlorine storage	Proximity of storage to boundary fence	Guard patrols	Chlorine release	Multiple fatalities on site and off-site	Consider fencing tank farm and providing intrusion detection system

3.5.4 Combined studies

Asset based and scenario based SVA have been integrated into a unified approach so that it is possible to conduct the simpler, asset based approach first and, if needed, transition smoothly into a scenario based analysis, either for the entire facility or parts of it that would be of benefit in the opinion of the analysts [19]. It is also possible to go directly to a scenario based analysis. Physical and cyber threats can be addressed in separate SVA studies or in an integrated study [19].

Tab. 3.6: Example of sneak path cyber SVA

ATTACKERS	TARGETS	PATHS	BARRIERS	EVENTS	CONSEQUENCES	RECOMMENDATIONS
Hacker	Reactor temperature control set points	Internet connection to site LAN	LAN frewall	Runaway reaction	On-site fatalities	Consider intrusion detection system
		Dialup modem on PC connected to PCN	Password			Consider elimination of PC
		Contractor network and use of dialup modem connection to site LAN	PCN firewall			Secure modem
Disgruntled operator	Tank farm control valves	HMIs	Fellow operators Alarms	Spill to dike	On-site fatalities	Restrict access to HMIs
		Desktop PC	Password			Consider elimination of PC
		EWS in engineer's office	Password			Restrict access to EWS

3.6 Stages and steps in SVA

This section describes the steps followed in performing SVA studies. Performance of SVA involves:

1. Preparation and organization
   - facility description
   - threat intelligence
   - team selection
   - study, purpose, scope, and objectives
   - subdivision of system
   - schedule and facilities
   - recording study results
   - quality assurance
   - communication with management
   - legal counsel
2. Target analysis
3. Threat analysis
4. Identification of vulnerabilities
5. Identification of consequences
6. Identification of existing countermeasures
7. Identification of enablers
8. Identification of recommendations
9. Documentation and reporting
10. Follow up

It should be noted that the step involving the identification of vulnerabilities often is referred to as *vulnerability analysis*, even though the term *security vulnerability analysis* is used to describe the complete set of steps listed. Each of these steps is described below.

Step 1 Preparation and organization

Facility description: Various types of information are needed to conduct an SVA. Included are information on chemicals handled and their properties, locations, and uses; a chemical reactivity matrix; equipment and materials used and their characteristics; recipes for batch processes; drawings such as piping and instrumentation diagrams (P&IDs), process flow drawings (PFDs), and plot plans; information on computer systems and utilities; and countermeasures in place. Information on the community, population, environment, neighboring facilities, and physical surroundings of the plant is also needed.

Documentation on existing safety and security programs will need to be consulted. Results from previous safety or security studies and reports on previous safety or security incidents should be reviewed. The occurrence of previous incidents must be reflected in the analysis. Also, when an incident occurs, the SVA should be reviewed to determine to what extent the incident was addressed. Insights obtained should be used to improve the quality of future SVAs. Compiled information must be accurate and up to date.

Various types of additional information are needed to conduct a cyber SVA including computer system architectures; network configurations; interfaces between systems and networks, internally and externally; security measures; system design and operation; control software logic; hardware and software used (operating systems, firmware, applications); and support systems and utilities. Automated scanning tools can be used to develop a profile of a computer system, for example, a network map.

This information needs to be gathered and, in some cases, prepared, if it does not already exist. Information gathering may involve administering questionnaires, collecting and reviewing written documents, conducting surveys, touring the facility and making observations, and interviewing facility personnel. SVA team members also contribute their knowledge of the facility during the performance of the SVA.

Threat intelligence. Threat analysis requires information, or *intelligence*, on threats including identifying possible adversaries and their motivations, intents, capabilities, characteristics, and tactics. Motivation is the reason for an attack, for example, to make a political statement or financial gain. Intent combines the objective of an attack and the means of achieving it using an asset, for example, the desire to cause public fatalities by causing a release of hazardous material. Capability refers to adversaries with both the ability to access an asset and to use it to achieve their intent. Characteristics relate to what adversaries are willing to do to accomplish their objectives, for example, sacrifice their own lives or those of others. Tactics relate to how adversaries will overcome countermeasures and exploit vulnerabilities to achieve their intent. Companies should use whatever information they have available but will need to consult with local, regional, and national law enforcement authorities, government agencies, industry organizations, community groups, industrial neighbors, and web based sources. Any history of system break ins, security violations, or incidents should be reviewed by consulting with facility personnel and reviewing reports. Knowledge of intrusions at other facilities in the area is also valuable.

Team selection. A multi-disciplinary team is needed that is capable of brainstorming threat events or scenarios and providing the perspective needed to adequately analyze security threats. One team member should be designated as the leader, or facilitator, and must be knowledgeable and experienced in the use of SVA, have an understanding of the types of security issues to be addressed, and possess team facilitation skills. The team leader should be impartial with no predispositions towards the outcome of the study.

Team members should be selected who together have appropriate knowledge and experience of the facility/process/equipment design, engineering, operation, maintenance, and layout; its security, safety, health and environmental features, methods, systems, procedures and programs; processes and their chemistry and controls; computer systems; materials handled and their physical, chemical, and hazardous properties and locations; equipment used and specifications; site characteristics; potential adversaries; and countermeasures including strategies for their use. It can be valuable to have at least one team member who does not work in the facility and can provide an outsider's perspective. Teams must be able to work together effectively. Team members should have skills or training in SVA.

For cyber SVA, team members should include people who are knowledgeable in the computer systems and computer support systems used, including their functions, operation, hardware, software, and peopleware; the topology, structure, and interfaces of networks; cyber vulnerabilities; techniques and tactics used by hackers and other attackers; and cyber security countermeasures.

Team members should be knowledgeable of actual plant operation, maintenance, safety, and security practices, since they may differ from written requirements. Similarly, team members should be sufficiently knowledgeable to be able to recognize when drawings or other documents contain inaccuracies. There is little point in spending the time and effort required to conduct an SVA if it is performed for a facility that exists only on paper.

Typical teams may have up to six or more members. The team should be large enough to brainstorm effectively and to provide all the knowledge needed, but it should not be so large that brainstorming is hindered. Usually, fewer than 3 or more than 8 people can create problems. The actual team size depends on the complexity of the facility and the expertise of individual team members. Some team members may provide knowledge and expertise in more than one area.

Study purpose, scope, and objectives. The study purpose defines the reason the study is being performed, for example, to protect against terrorism and to comply with industry requirements such as the ACC Security Code [1]. Generally, SVAs are conducted to protect people, property, the environment, the company, and national interests. The definition of purpose helps to ensure that the needs of the company, its employees, the public, and other parties with vested interests are met. The study scope identifies the attackers and intents to be considered, and the facilities, assets, and operations that are subject to these threats and are to be addressed in the study. The study objectives define the types of consequences to be included. Consequences are the adverse impacts resulting from an attack against an asset, wherever they occur, local or plant wide, within the facility, or externally. Impacts may be human fatalities and injuries; facility damage or loss; disruption of operations, the community, society, or the economy; environmental damage; and financial loss. They also may include loss of critical data, information, reputation, morale, and public confidence. Any known exclusions from the scope and objectives of the study or assumptions that will be made during the study must be stated and justified.

The statement of purpose, scope, and objectives helps to ensure a focused study that addresses all appropriate issues. It helps avoid digressions during the performance of the study.

Subdivision of system. A facility or process can be considered as a single system, or it can be subdivided into sectors for a more detailed analysis of threats. Subdivision helps to focus the analysis and is used to provide an appropriate level of detail consistent with the purpose, scope, and objectives of the SVA. For example, a physical SVA may consider a tank farm, production unit, and product storage area in a facility.

A networked computer system can be examined in its entirety as a single system or it can be broken down into sectors for more detailed analysis. The latter approach is preferred for situations involving complex and/or multiple networks. Sectors may be individual networks, for example, enterprise, process information, process control, etc. Large networks may be studied as subnets. Process control networks are usually studied for each individual process. Sectors may be organized according to the technical expertise needed to address each sector, for example, network engineers for IT security and process control engineers for control system security. Both physical and cyber-attacks on computer systems are covered in the sectors representing the computer systems.

Whenever subdivision is employed, a global sector should be used to account for threat events or scenarios that arise within multiple sectors and/or affect the entire facility/process/computer network. For example, adversaries may attack two different chemical storage areas simultaneously and such scenarios may not be identified if the storage areas are considered only within separate sectors. Similarly, attacks may have adverse impacts beyond the sector where they are initially made, and it is important that all impacts be identified. For a cyber SVA, adversaries may attack two different networks simultaneously and such vulnerabilities may not be identified if the networks are considered only within separate sectors. Similarly, attacks against a control system may have adverse impacts beyond the sector where they are initially considered, and it is important that all impacts be identified. Documentation of the analysis is provided for each sector when subdivision is used.

Sometimes SVAs may be leveraged. Once one type of system has been analyzed, it may be possible to edit it to create SVAs for other similar systems. For example, different computer control systems used to operate different processes within a plant will share some characteristics, for example, their connection to the same business network and operation by the same personnel. An SVA for the first system may be used as the starting point for the second system. Also, when one type of computer system has been analyzed, it may be possible to use it as a surrogate for other similar systems, as long as the results of the analysis will be similar. Leveraging SVAs in this way reduces the overall effort required for facility SVAs.

Schedule and facilities. The time required for a study should be estimated and team sessions scheduled. The time required will depend on the complexity of the facility, the threats faced, and the experience of the team with SVA. The schedule should be set as soon as possible to help team members plan for the time commitment required. A suitable meeting room must be arranged with the necessary equipment to perform the study.

Recording study results. A means must be made available to capture and record in written form the results of the SVA, including threats, assets, vulnerabilities, existing countermeasures, and recommendations for new or improved countermeasures. A team member, possibly the facilitator, should record or scribe the study as it is performed. Software tools may be used for this purpose.

Quality assurance. Provisions should be made to ensure a quality study is performed. This includes ensuring that information used is accurate and complete, the right team members are assigned, appropriate facilities are provided for the study, the facilitator is suitably qualified, and the study is technically correct. Arrangements should be made for review of the completed analysis by an experienced independent analyst.

Communication with management. Communication with management and others with vested interests is needed prior, during, and after a study. Prior to a study, management approval should be obtained for the participants and the schedule. Participants will likely report to various managers. While the statement of purpose, scope, and objectives often is prepared by the SVA team leader, it is actually the responsibility of management and their approval must be obtained prior to embarking on the performance of the SVA. During the study, management should be kept abreast of its progress and any results that may require immediate action. After the study, the results must be communicated to management for appropriate action.

Legal counsel. A number of issues may require legal guidance. This includes the content of study records and documentation, wording of reports and recommendations, confidentiality of information, and contractual terms for use of third parties.

Step 2 Target analysis

Target analysis is used to identify and screen possible assets for consideration. It involves:

- Identifying critical assets within the facility that may be attacked (*critical* means the assets, if attacked, could be used to cause harm).
- Estimating the likelihood that a facility will be targeted, that is, selected by an adversary for attack.

2.1 Identifying critical assets

Critical assets within the scope of the study and that are at risk must be identified. There are two key questions that must be addressed to determine whether assets are critical:

- Do they have attributes that enable their use to cause harm?
  For some assets, a single attribute may be important, while, for others, multiple attributes must be considered. For example, for people the intrinsic value of human life is the key attribute, while for a chemical it may be toxicity together with properties and a physical form that allows its easy dispersion. Hazardous materials by their nature can cause harm if released from their containment. However, assets need not be inherently hazardous to enable them to be used to cause harm. This is particularly true for cyber assets. It is through the manipu-

lation, disablement, or damage of cyber assets or the theft of information from them that harm is caused. Attributes for cyber systems include their financial value, stored data and information, and potential for manipulation or shutdown. Attributes for information include competitor value, cost to reproduce, and utility to an attacker. Attributes may be intangible, for example, the reputation of a company. Some attributes for common assets are provided in Table 3.7.

Tab. 3.7: Some attributes for common assets

Asset	Attributes
Chemicals	Hazardous properties such as toxicity, flammability, explosivity, corrosivity, and carcinogenicity
	Physical properties such as vapor pressure and boiling point
	Form such as liquid, gas, or pressurized liquid
	Concentration
	Quantity
	Location, such as proximity to the plant fence or occupied buildings on-site
	Thermal and chemical stability
	End use such as products used in food/nutritional supplement production, the manufacture of pharmaceuticals or cosmetics, or that are key to the economic viability of the company or nation
Equipment	Financial value
	Location
	Size
	Contents
	Construction
	Design
	Specifications
	Potential for misuse
People	Inherent value of human life
Computer systems	Financial value
	Stored data and information
	Potential for manipulation
Process and safety control systems	Financial value
	Potential for manipulation
	Potential for shutdown
Information	Competitor value
	Cost to reproduce
	Utility to an adversary

– Can *serious* harm be done?
  The judgment of what is serious needs to be made by each company. Typically, in SVA, it is the possibility of catastrophic impacts that is of concern. If no such attributes exist, or no serious harm can be done, the asset is not considered to be critical. For example, there may be no way in which an inventory of non-flammable oil could be used by an attacker to cause harm at a facility. However, if the oil were flammable, then it could be used to cause a fire. Care must be exercised in

making these judgments. For example, if the oil is used to make food products, then it could be contaminated by an attacker with possibly catastrophic results for consumers and the public when it enters the food supply.

Typically, information on critical assets is tabulated (Table 3.8).

Assets may be grouped for analysis by classes to provide structure to the SVA, for example, chemicals, computer systems, equipment, etc. Cyber assets may be grouped as hardware, software and data. Some of these classes may be further subdivided into categories, for example, chemicals may be divided into inhalation poisons, chemical weapons precursors, reactives, flammables, explosives, etc. This classification and categorization may help with decisions on countermeasures since different categories of assets may merit different countermeasure strategies. For example, protection against release of inhalation poisons will likely be different from the protection needed against diversion of chemical weapons precursors. Similarly, protection against destruction of computer hardware will be different from protection against intrusion into software applications and data bases.

Assets also may be grouped according to the type of threat to which they are most susceptible. For example, some chemicals may be targeted for release on-site owing to the potential for large-scale off-site damage while other chemicals may be targeted for theft for release at a later time and in a different place. Similarly, some cyber assets may be targeted for physical attack while others may be targeted for cyber intrusion and process manipulation.

A criticality factor can be used to rank critical assets according to their potential for causing harm. This provides a prioritized list for further attention, or allows the selection of specific assets that merit further analysis. In practice, it is often preferable to assign a simple prioritization factor based on judgement. Typically, a three, four, or five-point scale is used, such as very high, high, medium, moderate, and low.

Tab. 3.8: Example of target analysis

ASSETS	LOCATIONS	ATTRIBUTES	PRIORITY
Chlorine	Tank farm	Toxicity	High
Ammonia	Storage bullet	Toxicity Explosivity Ingredient for illicit drug manufacture	Medium Low Medium
Computer control network	"A" plant	Process control	High

2.2 Estimating likelihood of attack

Likelihood depends on the attractiveness of a target including the potential harm that can be achieved, ease of access to assets, and the difficulty of mounting a successful attack. Many factors are involved. Examples of are provided in Table 3.9.

Tab. 3.9: Examples of factors for target likelihood estimation

Item	Factors
Materials	Types of chemicals: – hazardous properties – environmental fate – physical properties – released form – exposure routes – ease of mitigation – breakdown products Inventories present: – amounts needed to be dangerous – proximity of storage containers Stored forms: – pressurized – liquefied On-site duration Number of rail cars, tank trucks and barges
Facility	Visibility: – visual from roads – public knowledge – Internet Appearance: – emblems – logos – signs – labels Recognizable as handling chemicals: – visible fractionation columns, storage tanks and other process equipment – presence of rail cars, tank trucks, and barges Layout: – proximity of assets to the plant boundary Location with proximity to: – population centers – critical infrastructure such as transportation centers, tunnels, bridges, power plants, water treatment plants, airports, ports, major highways, etc.; – other facilities subject to targeting – surface water and aquifers – provocative location Access: – barriers – manning levels – plant surroundings – intruders able to be observed – rail lines and roads (paved and unpaved including access and fire roads) Egress: – escape routes Importance of products: – sole supplier – tight markets

Tab. 3.9: (continued) Examples of factors for target likelihood estimation

Item	**Factors**
	Availability of information: – web sites – government filings – employee access Existing safeguards and secureguards Economic value Presence of multiple critical assets Importance to national and public interests
Surroundings	Topography: – channel a release – make concealment, intrusion and/or escape easier or more difficult Proximity to national assets or landmarks Meteorology to aggravate a release
Personnel	Operating hours: 24-hour operations are more secure Staffing level: presence of employees in sensitive areas Security personnel: – presence – visibility and numbers
Processes and storage	Production schedules: routines, advanced schedules and predictability facilitates planning for attacks Storage and processing time: the longer chemicals are in a hazardous state, the greater the window of opportunity for attack Frequency of use: for example, some batch processes may be run a limited number of times each year Location: indoors versus outdoors Types, sizes, numbers and construction of chemical containers Marking and labeling of vessels, tanks and lines Piping runs: longer lengths present greater exposure and more access points Building design: windows are vulnerable
Company	Prominence, influence, reputation, branding and public exposure: – a profile that makes it known to attackers – may be perceived to be capable of influencing the actions of government or others Connection with the government: – government-related work – products produced for the government Symbolic value Economic impact of loss of production

Tab. 3.9: (continued) Examples of factors for target likelihood estimation

Item	Factors
Community	Facility and community response and law enforcement capabilities: – availability – response time – staffing levels – equipment and training Emergency medical treatment: – availability – response time – capacity – proximity Potential for exposure and publicity in the media Opportunity for adversaries to convey their motive or message Level of hostile activity: history at facility, in the area, the industry and the nation

In assessing such factors, key issues are the potential for catastrophic effects such as mass casualties/fatalities/terror, extensive property damage, disruption to a country's infrastructure, and serious financial, economic, or environmental impacts.

Facilities may be targeted because they provide multiple assets of interest to adversaries. A facility offering multiple targets and opportunities to cause harm can be more attractive to attackers. For this reason, the overall attractiveness of a facility should be assessed before the likelihood of attack on individual assets is considered. A low facility attractiveness may obviate the need for detailed analysis of the likelihood of attack on individual assets.

Various approaches have been developed to estimate attack likelihoods including the use of judgement and ratings schemes [2, 9]. They may be useful when considering comparisons between facilities and to monitor changes at specific facilities. Likelihood estimation can be performed for a facility, an individual process, specific assets, or for each individual type of threat. However, all these estimates are necessarily subjective.

Target analysis results in a list of critical assets that is carried forward to the next step of the SVA, threat analysis. Threat analysis may be conducted only on assets that exceed priority levels. For example, in the target analysis shown in Table 3.8, only those assets which are of "high" or "medium" priority may be carried forward. "Low" priority assets may not be considered further.

Some organizations have developed simple screening approaches to prioritize facilities for analysis, for example, the ACC in its Security Code [1]. A modification of the ACC screening scheme has been applied to cyber security [19].

Step 3 Threat analysis

Threat analysis involves the identification of attackers, their intents, and the criticality of credible threats [9]. Credible threats are ones believed to be possible. Determination of credibility involves subjective judgment but, given the unpredictable nature of malevents, it is wise to err on the side of conservatism and include threats that may be considered of very low likelihood. This provides the benefit of evaluating and recording such threats and at least considering countermeasures that may protect against them. Furthermore, credit can be taken for their low likelihood in the estimation of criticality.

Systems are analyzed to identify ways that attackers could combine with targeted assets to accomplish the intent of the attackers and cause harm. The pairing of attackers and their intents with assets identifies threat events. Threat events may be screened for consideration in vulnerability analysis. Some pairings may not be considered likely, for example, terrorist targeting of an inventory of potassium cyanide (a solid material) for release on-site may not be judged as likely as its theft for use in contaminating food and water supplies. Threat analysis produces a list of threats that are considered credible.

Threat analysis involves:

- identifying potential adversaries with the desire to cause harm;
- identifying the intent of adversaries;
- assessing the criticality of the threats.

3.1 Identifying potential adversaries with the desire to cause harm

Systems are subject to attack by various adversaries, including the following:

- Terrorists and saboteurs. Process facilities can be attractive targets for politically motivated groups.
- Disgruntled insiders who damage systems, steal information, or cause other harm for revenge, profit, or other purposes.
- Thrill-seeking hobbyists or alienated hackers who gain a sense of power, control, accomplishment, self-importance, and pleasure through successful penetration of computer systems to steal or destroy information or disrupt an organization's activities. Often, they are motivated by the fame and notoriety they gain in the community of hackers and/or the media.
- Professional thieves who steal information or other assets for sale.
- Adversary nations.

Key threats can be identified by reviewing checklists of potential attackers and considering available information on current threats. A decision flowchart has been proposed for use in threat assessment [54]. Threat analysis is a subjective process and no listing of potential attackers is ever likely to be complete.

3.2 Identifying the intent of adversaries

Adversaries may have the objective of causing harm to employees, the public, the company, a facility, an industry, the economy, national security, etc. They do so through such means as the release of hazardous chemicals and the shutdown of a process. Intent is the combination of the objective of an attack and the means of achieving it using an asset.

Intents to consider for physical attacks include:
- causing the release of hazardous materials;
- theft/diversion of assets for use in causing harm;
- contaminating products to cause harm;
- damaging, destroying, or stealing assets;
- manipulating or disabling assets to cause harm;

Intents to consider for cyber attacks include:
- Manipulation of cyber assets to cause a hazardous material release, runaway reaction, diversion of materials for use in causing harm, contaminating, or poisoning products, etc., for example, hacking, physical attack, unauthorized operation.
- Disablement, damage, or destruction of cyber assets to prevent their proper operation or cause a financial loss, for example, physical attack, cutting cables, denial-of-service attack, malware.
- Loss, theft, disclosure, damage, destruction, or corruption of data or information stored in cyber assets, for example, hacking, theft of storage media, and portable computers.

Industrial cyber security goes beyond considering just data or information assets, as is typically done in IT cyber security which addresses the integrity, availability, and confidentiality of data and information. Industrial cyber security also addresses other ways in which cyber assets can be used to cause harm.

3.3 Assessing the criticality of the threats

Sometimes threat analysis includes estimating the criticality (likelihood and/or severity) of specific threats in order to prioritize or select them for consideration during the identification of vulnerabilities. However, severity estimation at this stage is of questionable value, since it is difficult to prioritize threats with dissimilar intents that range from releasing chemicals to contaminating products. In some cases, analysts may treat all threats equally from the perspective of their consequences and prioritize them on the basis of their likelihood only.

Estimation of threat likelihood is a refinement of the estimation of the likelihood the facility will be targeted (see Step 2) to incorporate the likelihood of a specific attack. Threat likelihoods represent beliefs regarding the choices that adversaries may make. Threat likelihood estimates can be made as part of the threat analysis or when the overall likelihood of a threat scenario is estimated after vulnerabilities have been identified and existing countermeasures and possible consequences have been considered. In the former case, a simple criticality ranking (for example, high, medium, low) may be used as a composite measure of likelihood and severity. However, likelihood estimation for malevents poses challenges (see Section 3.9, Beyond SVA).

Formal threat ranking schemes have been developed for process facilities [35]. However, an assumption simply may be made that possible threats exist so that resources can be focused on vulnerability analysis.

It is important to understand how motivation relates to targets and to consider the capabilities of adversaries [32]. There are various motivations for threats. They include political, social, issue-oriented, religious, ideological, economic, and revenge/retribution. The scope of credible threats can be narrowed by considering possible motivations of potential adversaries to determine if they will result in a specific company or facility being targeted. It is important to try to look at the company through the eyes of adversaries when doing so. Intelligence on potential adversaries is vital to this analysis. Correlations of motivations, intents and targets help in identifying threats. Adversaries may be motivated but not capable. Capabilities include training, equipment and know-how. It is important to make conservative assumptions with regard to capabilities because often "where there is a will, there is a way". Adversaries may enlist the assistance of technically qualified people who may participate either knowingly or unknowingly.

The results of the threat analysis are recorded in a spreadsheet (see example in Table 3.10). The threat events identified in threat analysis are studied when addressing vulnerabilities. Threat analysis identifies what could happen while vulnerability analysis identifies how it might happen.

Tab. 3.10: Example of threat analysis

ASSETS	ATTACKERS	INTENTS	CRITICALITY
Chlorine	Disgruntled employee Terrorist	Release Release	High Medium
Ammonia	Disgruntled employee Terrorist Drug trafficker	Release Release Theft	High Medium Low
Computer control network	Hacker Contractor	Shutdown process Environmental release	Medium Low
Food oils	Activist	Contaminate foods	Medium

Step 4 Identification of vulnerabilities

Vulnerability analysis identifies ways in which the threat events from threat analysis can be realized (i. e. threat scenarios) by brainstorming in a similar way to identifying hazard scenarios in PHA in order to identify weaknesses that can be exploited by an adversary to gain access to critical assets in order to exploit them. Brainstorming focuses on the penetration and action elements of threat scenarios (Figure 3.4). SVA teams identify how the system can be penetrated and what malicious actions can be taken once access has been gained. Brainstorming is intended to result in the discovery of less obvious vulnerabilities and produce a deeper understanding of how assets can be exploited. Teams need to guard against focusing attention only on predictable scenarios at the expense of less-obvious attacks. Threat criticality can be estimated for each individual threat scenario.

Addressing vulnerabilities for malevents is different than for accidents. Accidents result from random events (equipment failures, external events, or human errors). In contrast, malevents result from deliberate planned acts by attackers. The threats are intelligent in that attackers search for vulnerabilities, devise innovative attacks, and are capable of adapting to countermeasures and devising alternative strategies.

Vulnerabilities may be present in any part of a facility. Some examples are:

- Facilities, for example, poor fencing.
- Buildings, for example, lack of access controls.
- Processes, for example, accessibility of manual controls.
- Equipment, for example, manual valves that can be opened.
- People, for example, susceptibility to coercion.
- Location of people, materials, equipment and buildings, for example, located in remote area of site.
- Computer systems, for example, lack of intrusion detection.
- Utilities, for example, ease of access.
- Policies, for example, unescorted visitors allowed.
- Procedures, for example, no screening of delivery personnel.

Further examples of vulnerabilities are provided in Table 3.11. Vulnerabilities in countermeasures also must be addressed (Table 3.12).

Tab. 3.11: Examples of vulnerabilities

INTENTS	VULNERABILITIES
Release of hazardous chemical	Manual valves not locked
	Contents of vessels and lines clearly marked in plain language
	Manual overrides
	Unprotected vessels and piping subject to ramming by a vehicle
	Long accessible piping runs
Reactivity incident	Ability to add a contaminant to reactants
	Ability to alter process conditions, for example, temperature
	Ability to cause loss of agitation
	Ability to disable emergency shutdown
Shut down production	Insecure utilities
	Access to manual overrides
	Access to emergency shutdown
Contamination of products	Ability to add contaminants to process chemicals
	Insecure storage of products

Theft of chemicals	Poor employee and contractor screening No vetting of carriers Lack of supervision Material is stored in small containers Waste or rework material is produced Samples can be taken Storage containers are not sealed Tamper-evident storage is not used No material accountability or tolerances are large Warehousing and storage areas are not secure

Tab. 3.12: Examples of weaknesses in countermeasures

Countermeasure	Weakness
Inherently secure technologies	Designs and measures not used
Layout	Sensitive areas close to the facility perimeter
Passive safeguards, e. g. dikes and barriers	Lack of testing and maintenance
Active safeguards	Common-cause failure from deliberate acts
Emergency shutdown procedures	Do not address malevents
Fences	Not high enough Not protected with topguard Gaps or holes Deter vandals only Adjacent structures help overcome
Gates and doors	Propped open for convenience Railroad gates inadequate
Vehicles	Failing to check underneath Failing to check cabs
Contractor screening	No auditing of the contractor security program
Access to process areas	Not restricted
Document control and safeguarding	Not used or enforced
Cyber security	Few or weak countermeasures taken
Emergency response	Plan does not address malevents Inability of responders to deal with anti-personnel devices Law enforcement inadequately equipped Response times too long Fire department overwhelmed Drills not conducted for malevents Insufficient PPE for malevents No crisis management plan

The identification of physical security vulnerabilities often involves an examination of the actual facility. Similarly, cyber vulnerabilities require an examination of the actual computer systems. However, this requires the use of specialized methods to search for vulnerabilities that may not be known, for example, insecure modems and weak passwords. This should be done not only as part of cyber SVA but also on a regular basis as part of a cyber security program [18]. Penetration testing can be performed by "white-hat" hackers. Automated vulnerability scanning tools are available, although they can produce false positives. Security testing and evaluation also can be used to determine the efficacy of existing countermeasures.

All aspects of computer systems, hardware, software, data and peopleware, may contain vulnerabilities. Vulnerabilities of computer systems can be categorized as providing or facilitating access, or facilitating misuse. Attackers use a variety of techniques and tools to exploit these vulnerabilities including hacking software, reconnaissance, social engineering, password crackers, scanning, war dialing, sniffing, and spoofing [11]. Some cyber vulnerabilities will be known and they can be identified in discussions with system administrators, users, and support personnel. Known cyber vulnerabilities also can be identified by consulting industry sources such as websites of vendors where system bugs and flaws are listed together with bug fixes, service packs, patches, and other remedial measures and security advisories from government organizations and commercial organizations. Identification of other specific vulnerabilities depends on a knowledge of the types of cyber vulnerabilities possible and the ability of the SVA team to recognize them in the system being studied.

Checklists can be used to guide brainstorming of vulnerabilities. However, it is especially important to try to think "outside the box" when brainstorming vulnerabilities. Creative thinkers should be involved [27]. Adversaries often do not have the resources to mount military-style operations. Instead they use their time and energy to devise creative ways to attack. Their terrorist or criminal background does not mean they are not challenging adversaries. They must be cunning to overcome the obstacles that face them.

Traditionally, in the context of identifying scenarios, risk assessment has asked the question, what can go wrong? [47]. Kaplan observed that the TRIZ theory of problem solving incorporates anticipatory failure determination which asks instead, the question, if I wanted to make something go wrong, how could I do it? [46]. This rephrasing of the question may be particularly useful in the context of identifying threat scenarios.

SVA team members need to assume the mindset of an attacker. In considering how adversaries may exploit vulnerabilities, a perspective on what is credible should be maintained. While adversaries may be creative, they are still human and subject to human limitations on what they may accomplish. Keep in mind, however, that humans who are motivated in the extreme can accomplish unusual acts. It is also important to be aware of what security incidents have previously occurred, either at the facility or in industry generally, since they may provide insights into what is possible for the facility in the future.

Most situations involving deliberate acts are not likely to be simple. For example, domino and cascading effects may be triggered by attackers [48, 53]. Since the human mind knows no limitations, threat scenarios could be quite complex. However, there may be effectively an infinite number of possible threat scenarios and, by necessity, their analysis requires the consideration of representative scenarios that encompass the spectrum of possibilities.

Vulnerability analysis must consider possible tactics used by attackers. Some key issues to address include:

- What information may have been obtained to plan an attack?
- How easily can assets be identified?
- How can the facility and areas within it be penetrated (covertly, use of deceit, by force)?
- How can a malevent be caused?
- How much time will be available for an attack?
- Is an attack more likely to succeed at a particular time (of day, production step, lifecycle activity)?
- Can countermeasures be disabled?
- Can multiple countermeasures be disabled simultaneously?
- How can maximum consequences be produced?
- How can the facility response be neutralized or impaired?
- What weapons may be used?

In some asset based SVA methods, vulnerabilities are not recorded explicitly in the worksheet, nor are existing countermeasures [35]. Rather they are examined when recommendations for new or improved countermeasures are considered (Tables 3.1 and 3.2). However, the analysis is clearer if vulnerabilities and existing countermeasures are explicitly recorded. This practice can be contrasted with typical scenario based SVA where vulnerabilities and existing countermeasures usually are recorded (Tables 3.3 and 3.4). Sufficient detail must be provided to allow the scenarios to be understood by SVA reviewers.

One challenge for SVA is that adversaries are intelligent and adaptive. They may find vulnerabilities that have not been identified, and they are able to adjust to new security countermeasures in planning an attack and modify their actions in real time during an attack. Moreover, the intents, capabilities, motivations, characteristics, and tactics of adversaries may change over time for reasons not associated with the facility targeted. Thus, unlike a hazard scenario in safety risk assessment, threat scenarios are dynamic. In a hazard scenario, an initiating event such as the failure of a basic process control system to control the level in a tank results in process safeguards being challenged. Safeguards may include, for example, a high level shutdown system, a dike (bund), and a deluge system. The scenario can be modeled by considering the success or failure of the safeguards to operate at each point in the scenario. However, in a threat scenario, an intelligent adversary who gains knowledge of countermeasures that have been installed as the result of a security risk assessment has the opportunity to adapt their tactics. Furthermore, attackers will try to adapt during an attack as each countermeasure is encountered and it is challenging to predict what adaptations may occur. SVA teams should consider possible adaptive responses by attackers and factor them into their analysis.

Step 5 Identification of consequences

Possible events of concern resulting from attackers reaching targets using vulnerabilities are identified together with their potential impacts. Types of consequences to be considered may include employee or public fatalities and injuries, environmental damage, property damage, financial loss, loss of production, loss of critical information, disruption of company operations, destruction or disruption of critical infrastructure, loss of reputation, etc. Collateral damage to adjacent facilities may be an intended or unintended consequence of an attack against a facility. Consequences are recorded in the SVA worksheet (see examples in Tables 3.1–3.6).

Consequences may be specified in more detail than their overall impacts, for example, a simple measure for chemical releases is the quantity of material released. More specific measures can be used, for example, the impacts of a hazardous material release could be expressed as the distance to a "level-of-concern" endpoint concentration for toxics, an explosion overpressure value for explosives, and a thermal radiation level for flammables. It is also possible to calculate actual impacts on people using dispersion analysis and dose-response modeling. For the consequence of plant shutdown, the measure may be the financial value of lost production. However, it is usually desirable to avoid quantitative analysis in SVA. Consequently, often the type of consequence is recorded and its severity may be estimated as part of a ranking scheme.

Usually, a range of consequences will be possible for each threat event or scenario. The worst-case consequence should be assumed to be conservative. There may be multiple layers or rings of countermeasures present. However, attackers may attempt to breach all existing countermeasures simultaneously. For example, the use of explosives or other means to disable or bypass countermeasures is likely, and the attackers may be willing to sacrifice their own lives to accomplish this objective. Thus, when assumptions are made, they should be conservative. Consequences of malevents may well be much more severe than accidents since attackers will be attempting to cause maximum damage. Furthermore, consideration must be given not only to attacks that may be completely successful but also those that may be partially successful.

Usually, it is acute health effects rather than chronic ones that are of interest in SVA because of their immediate impacts. However, adversaries may attack facilities that have materials of chronic toxicity because their release could cause panic and fear among the public. Therefore, consideration should be given to including such materials, as appropriate.

Step 6 Identification of existing countermeasures.

Existing measures that may counteract a threat, reduce or eliminate vulnerabilities, or mitigate consequences are identified so they can be considered when the need for new or improved countermeasures is discussed. In asset based SVA, countermeasures usually are considered when recommendations for new or improved countermeasures are discussed and typically are not recorded in the worksheet. In contrast, countermeasures are recorded in the worksheet for scenario based SVA (Tables 3.3 and 3.4).

Countermeasures may address prevention, detection, control, and mitigation of threat events and scenarios. They may be safeguards or secureguards. Safeguards often are layered and secureguards may be arranged in rings of protection. Failure of these safeguards and secureguards increases the likelihood of a successful attack. A classification scheme has been described for secureguards and safeguards to protect against deliberate acts [8].

When considering the applicability of existing countermeasures, the following issues should be addressed:
- Functionality, i. e. Is it bypassed, disabled or removed?
- Adequacy, i. e. Is it enough?
- Applicability, i. e. Does it really apply? Is it directly applicable?
- Effectiveness, i. e. Does it accomplish its purpose?
- Reliability, i. e. Will it work?

When considering the effectiveness of existing countermeasures, attacks that disable them should be considered. For example, attackers may disable a safety instrumented system before causing a hazardous chemical release or initiating a runaway reaction. Actions of attackers may deliberately or inadvertently cause common-cause or common-mode failures that can compound an attack. For example, if the use of explosives disables a main computer system, control of many process functions may be lost.

Safeguards against accidents may well be inadequate for malevents. Typically, accidents have lesser consequences than malevents, and safeguards designed to dump chemicals to a quench tank, flare, scrubbing system, another means of disposal, or backup storage likely will be overwhelmed. Moreover, they may be disabled by attackers during a malevent.

Step 7 Identification of Enablers

Enablers are events or conditions that must be present or active for a threat event or scenario to proceed. They do not initiate a threat event or scenario by themselves but make them possible. Also, they impact the likelihood of threat events and scenarios. Enablers may influence any of the elements of threat events and scenarios. Examples are:

- Availability of facility information
- Prevailing meteorological conditions
- Unattended operation

Such factors should be considered in SVA and may be recorded in the SVA worksheet.

Step 8 Identification of Recommendations

Recommendations may be made for new countermeasures or enhancements to existing countermeasures. The need for additional or improved countermeasures is determined based on the possible consequences, existing countermeasures, vulnerabilities, nature of the threat, and the risk reduction afforded by the proposed countermeasures. Teams need to judge whether or not recommended countermeasures are sufficient to reduce the threat risk to a tolerable or acceptable level.

Specific guidance can be provided on tolerable or acceptable risk levels, as is sometimes done for accident risk [36]. However, the application of the concept of risk to malevents is controversial (see Section 3.9, Beyond SVA).

It is also possible to define security performance standards according to threat type. For example, one set of specific countermeasures may be required for the threat of hazardous material release, versus a different set for the threat of diversion of chemicals. Another approach is to protect assets according to the highest-level threat to the asset [35]. This is sometimes done in asset based methods. However, it can lead to unprotected vulnerabilities since protection against one threat, no matter how high its risk, may not provide protection against lower risk, but still significant, threats. A preferred approach for asset based studies is to consider countermeasures for each threat event. This requires a little more work but helps provide assurance that countermeasures have not been overlooked. In the case of the more detailed scenario based analysis, countermeasures are considered for each scenario.

Various types of countermeasures are possible [39]. It is desirable to pursue an overall strategy for countermeasures [31] and countermeasures must be part of a security program for a facility [8]. A hierarchy of countermeasures can be established:
- Prevention
- Detection
- Control
- Mitigation

Prevention includes physical security such as access control and barriers, information security such as document control, computer security such as firewalls, and inventory control. *Detection* includes surveillance, intruder detection, alarms, and monitoring for the presence of chemicals or process parameters such as flow or level that may indicate a release. *Control* uses facility layout, measures to limit releases in the event containment is lost, and emergency shutdown in the event of an attack. *Mitigation* can include stockpiling chemical antidotes; the use of engineered safeguards such as projectile shields and containment structures; and emergency and law enforcement response.

A hierarchy of security measures can also be followed:
- Make assets less attractive, for example, change their location.
- Eliminate or reduce the threat, for example, restrict control room access to operators.
- Eliminate vulnerabilities, for example, eliminate Internet connection to a control system.
- Provide layers of protection, for example, authentication plus firewall plus intrusion detection.

Inherently securer and safer technologies also should be considered. A set of inherently securer technology principles has been proposed [13]. The goal is to produce a facility that is "benign by design" by eliminating or reducing features that make the process attractive to adversaries. Inherently securer/safer approaches reduce or eliminate risks in ways that are permanent and inseparable from the design. However, other protection layers also should be provided, including an emergency response program that addresses threats. Although inherently securer/safer approaches are best considered during the design of new processes, it is also possible to retrofit some features for existing facilities.

Process design also offers the opportunity to consider equipment that is appropriately resistant to attack, for example, increased wall thicknesses, double-walled construction, mounding, and underground installation. Where possible, weak points such as sight glasses and flex hoses should be avoided. Protection for critical support systems such as computers, utilities, and communications should also be addressed during design, for example, placing wiring in rigid conduits.

The layout of equipment and buildings is part of the design process for a new facility. Generally, hazardous materials and sensitive areas should be located away from the facility perimeter for improved security. The most vulnerable locations should be the hardest for adversaries to reach. Sensitive areas include control rooms, computer rooms, motor control centers, rack rooms, server rooms, telecommunication rooms, and utilities.

In recommending countermeasures, it is useful to consider the application of some traditional security and safety philosophies. For example, the defense-in-depth concept is based on the premise that multiple layers or rings of protection ensure some level of protection in the event that one or more layers or rings fail. Another concept that is important for process security is the use of both high and low profile systems. High profile systems are intended to be noticed by and discourage adversaries, while low profile systems provide protection against determined adversaries who are not discouraged by the high profile systems but may not readily detect the low profile systems. A further process security concept is to ensure an appropriate balance between secureguards and safeguards to provide diversity and more reliable security and safety. For cyber security, the principles of separation of functions, isolation, need-to-know, and least access are important [11]. However, both the advantages and disadvantages of the application of these philosophies for malevents must be understood. For example, the classical asset based security philosophy of deter, detect, and delay is seriously flawed for terrorist physical attacks against plants, but has merits for cyber threats. The approach works well for the protection of assets such as valuables in a bank vault. However, when the assets being protected are hazardous chemicals and the adversaries are terrorists, the approach is of limited utility because if deterrence fails and adversaries are detected and delayed, it assumes a response force will have sufficient time to respond and take appropriate action. Unfortunately, in the case of hazardous materials and terrorists, response times may not be fast enough to prevent terrorists from taking their intended actions. Also, the ability of typical response teams to neutralize a group of determined, armed, and equipped terrorists is questionable.

Process facilities may gain some protection from black and gray swan events[1], cascade failures, domino events, dependent failures, and systematic failures by applying the "3 R's", resiliency, redundancy, and robustness, as a design philosophy [42]. Resiliency is the ability to recover from a disruptive event, redundancy is the provision of a process function in multiple ways, and robustness is the ability of a system to withstand deviations from normal conditions. The concept of resilience has been applied to chemical industrial areas [55].

Considerations when recommending countermeasures include adequacy, applicability, effectiveness, and reliability, as for existing countermeasures. Additional considerations for new or modified countermeasures include their impact on safety, operations, quality, or working conditions, i. e., do they impair operability, safety, quality, or ability to work? Actions to enhance security could adversely impact safety, operability, etc. For example, plants are often built out-of-doors so that leaks can be dispersed by natural ventilation. Enclosing them in buildings may provide more security by restricting visibility and access, but at the expense of increasing the risk of exposure to personnel within the building unless special precautions are taken. A number of companies have relocated control rooms away from process areas to improve process safety as part of facility siting studies. In some cases, these relocations may have resulted in a less secure facility. Malevents that involve opening valves to tanks and vessels can be mitigated by smaller pipe sizes, restriction orifices, and lower pump capacities, but this may conflict with production requirements. Process computer control networks are increasingly being interfaced with business and enterprise networks to provide business and operational efficiency but at the expense of increasing the risk of cyber penetration by attackers. Enhanced password protection using lockout after several logon attempts to

1 A black swan event is a rare, unpredictable, and catastrophic occurrence, and a gray swan event is one that was recognized but dismissed as not credible, but occurred [56].

improve security may not be possible for computer control systems for safety and/or operability reasons. Tradeoffs must be examined carefully in making such decisions.

Existing safeguards that protect against accidental releases may also protect against deliberate releases and diversions, but it is unlikely they will be sufficient. They may need to be strengthened, for example, automatic shutoff valves capable of being deliberately disabled may need to be replaced and additional safeguards and security measures may be required, for example, projectile shields to protect vessels from airborne and propelled explosive devices and projectiles.

With regard to countermeasures, disclosure of the existence of countermeasures may deter attacks but also aid in planning them. Decisions must be made regarding the disclosure of such information [57].

Step 9 Documentation and reporting

A written report is needed to facilitate review of the study, communicate its results to management, and assist in periodic revalidation of the SVA. It must be structured to meet the needs of different audiences including management and technical reviewers.

The report should describe the results of team deliberation, the SVA method used, how the study was performed and its technical basis. The report should also document information used; study purpose, scope and objectives; any risk estimation method employed, for example, a risk ranking scheme; assumptions made; and study participants with their areas of expertise. Results provided in the report should include the security vulnerabilities found and recommendations for new or improved countermeasures.

SVA worksheets usually are provided as a report appendix. Additional entries can be made in the worksheets beyond those shown in the examples. For example, category columns can be provided to categorize entries in other columns such as assets, attackers and their intents, vulnerabilities, consequences, countermeasures, and recommendations. Category columns are valuable for filtering and sorting information in worksheets, performing statistical analyses of the results, and generating customized reports. Other worksheet columns can be provided to track and manage recommendations including the assignment of responsibility, recommendation status, start and end dates, and comments on the resolution of recommendations.

The SVA worksheet examples provided use the column order: assets, attackers, intents, vulnerabilities, consequences, countermeasures, and recommendations. Alternative forms of the worksheet may be used in which the order of the first three columns is varied, according to the preferences of the analysts, for example, attackers, intents and assets. Also, it is possible to combine some of these items into a single column, for example, attackers and intents, although separate columns are desirable when attackers may have more than one intent. Alternative names are sometimes used for some of these items, for example:

asset: target
attacker: source
threat: attacker plus intent
vulnerability: path
countermeasure: barrier, safeguard, secureguard
consequence: event, impact

On completion of an SVA, management must be confident of the quality of the results. This depends on how the SVA was conducted and can be confirmed by a quality control review of the results. Such reviews are performed using criteria based on the method applied. Reviews can be incorporated into the performance of SVA or performed as part of report preparation.

Study documentation and reports contain highly sensitive information and must be controlled and safeguarded while still ensuring that the principles of community-right-to know and employee participation are met to the extent reasonable and appropriate.

Step 10 Follow up

Results of SVA studies must be communicated promptly to management for timely review and resolution of recommendations. Resolution may result in the adoption of recommendations for implementation as action items, modification of recommendations, the development of alternative recommendations, or the rejection of recommendations. The results and reasons for the resolutions of recommendations should be documented. In cases where recommendations are modified, substituted, or rejected, the result should be communicated to the SVA team to provide an opportunity for feedback to management. While it is management's prerogative to make the final decision on recommendations and the level of risk that is tolerable or acceptable, it is important that management fully understand the SVA team's intent for recommendations.

Issues for consideration in the review process are:

- How much risk reduction is provided?

  It is useful to prioritize action items according to the threats they ameliorate in order to assist in the allocation of resources. The entire set of recommended countermeasures must be considered to help ensure the residual risk to the facility is tolerable or acceptable.

- At what cost?

  Costs and benefits must be balanced, particularly with regard to the relative risk reduction provided by different countermeasures and the costs involved. Costs for countermeasures include costs of selection, procurement, purchase, installation, training, maintenance, additional personnel who may be needed, and adverse operational impacts of security measures. Once total costs have been estimated they should be factored into cost–benefit analysis to assist in selecting preferred countermeasures.

- Are there preferred alternatives?

  SVA teams may not recommend the most appropriate countermeasures. There may be other more effective measures available, lower cost measures that accomplish the same risk reduction, or measures that are preferred because they ameliorate more than one threat.

- Is the recommendation feasible?

  Countermeasures must be acceptable to affected parties for them to be successful. For example, placing locks on gates will be of little use if personnel leave them unlocked, and process operators may be unwilling to use passwords. Countermeasures also must be compatible with the existing facility. For example, setbacks cannot be provided if there is not sufficient space, and a new intrusion detection system may not be capable of implementation on a legacy computer system.

A goal of the review process is to try to ensure resources are applied where they will be most effective.

A tracking system is needed to help ensure recommendations are reviewed, resolved and, as appropriate, implemented. Responsibilities for the implementation of action items must be assigned, schedules established, and resources allocated to ensure their implementation.

SVA results should be communicated to affected people who need to know. For example, the security staff need to be informed of weaknesses identified and plans to correct them, operations personnel should be informed of changes that are planned to the process to improve security, IT managers should be informed of cyber vulnerabilities, and responders should be provided with information on the types of attack expected and their possible consequences.

3.7 Updating and revalidating SVA studies

Changes in process plants can occur frequently and threats may change even more rapidly. New cyber exploits are constantly being devised by attackers. These changes may affect threats and vulnerabilities. Thus, SVAs should be updated whenever any significant change occurs in the facility, the threats it faces, other aspects that may affect the risk, or after the occurrence of a security incident.

SVAs also should be revalidated on a regular schedule to ensure they reflect the current facility configuration, potential targets, and the present threats. Typically, revalidation involves reviewing the previous SVA to determine if any modifications are needed based on changes that have occurred to the process and the threats to which it is subject.

3.8 PHA and SVA

Some practitioners believe that PHA can simply be extended to include threat scenarios. However, threat scenarios are very different from hazard scenarios, and the latter are not a good basis for extrapolating to the former. The events that make up threat scenarios would be viewed as incredible in PHA studies and will usually have far greater consequence severities. For example, a hazard scenario may involve the failure of a drain valve on a tank in a tank farm causing a release. However, a threat scenario involving drain valves more likely would involve opening multiple valves in order to cause a much more serious release. PHA and SVA studies should be conducted separately.

3.9 Beyond SVA

Security risk practitioners often attempt to assess the risk of threat scenarios in a similar way to the assessment of the risks of hazard scenarios for accidents in order to help make decisions on the importance of threats, the need for additional or improved countermeasures, and prioritizing their implementation. In addressing the safety and security of critical systems, Aven noted that "we need to see beyond vulnerabilities and use a risk-informed approach" [4]. However, Guikema and Aven noted "assessing the uncertainties in and severity of the consequences of intelligent attacks are fundamentally different from risk assessment for accidental events and other phenomena with inherently random failures" [41]. Security risk is fundamentally different from accident risk and its treatment poses special challenges [30].

Process hazard analysis often incorporates the use of risk matrices to provide risk estimates for hazard scenarios [20]. Increasingly, layers of protection analysis (LOPA) is used to improve the scenario risk estimates [21]. Sometimes quantitative risk analysis (QRA) is employed [34]. Similar methods have been used to estimate security risks.

3.9.1 Risk matrices and risk scoring methods

Risk matrices are used in many technical areas including security [49]. They involve the assignment of severity and likelihood levels to estimate the risk of threat events or scenarios. The method relies on the use of engineering judgment and involves considerable subjectivity [25]. Team members may be tempted to shape risk estimates according to prejudices, biases, or desired outcomes. Furthermore, the use of risk matrices poses a number of challenges [22, 23, 33]. Their design and use requires considerable care [28,29].

Risk graphs pose similar problems and other risk scoring methods are problematic [43].

3.9.2 Rings of protection analysis

Recommendations for security improvements based on engineering judgment or risk matrices involve subjective judgment. They can produce inappropriate measures to reduce risk. LOPA is a simplified risk assessment method that is used to evaluate accident risks and provide more rational, objective, and reproducible decisions. A similar technique, rings of protection analysis (ROPA), has been developed for malevents [16]. Both security and safety programs typically use *defense in depth* to protect against threats and accidents. This is called *rings of protection* in security and *layers of protection* in safety. ROPA assists in identifying and determining the adequacy of existing protection systems. It is used to help determine whether there are sufficient rings/layers of protection against a threat scenario and whether the risk can be tolerated. A scenario may require multiple protection rings/layers depending on the process and the potential severity of the consequences. ROPA is intended to provide the basis for clear, functional specifications of required protection layers.

3.9.3 Security risk analysis

Some practitioners use QRA for security risks. QRA is based on the assumption that the events for which risk is quantified are stochastic in nature (i. e., random). It can be debated philosophically whether deliberate acts should be viewed stochastically or deterministically. This issue underlies all methods of assessing security risks. Notwithstanding this issue, available approaches for risk analysis of malevents have been reviewed and compared [41].

References

[1] ACC. Implementation Guide for Responsible Care® Security Code of Management Practices, Site Security and Verification, American Chemistry Council; July 2002.
[2] API. Security Vulnerability Assessment Methodology for the Petroleum and Petrochemical Industries. American Petroleum Institute; 2003.
[3] API. Security Risk Assessment Methodology for the Petroleum and Petrochemical Industries. American National Standards Institute/American Petroleum Institute Standard ANSI/API 780; 2013.
[4] Aven T. Identification of safety and security critical systems and activities. Reliab Eng Syst Safe. 2009;94;404–411.
[5] Bajpai S, Gupta JP. Site security for chemical process industries, J. Loss Prevent Proc Ind. 2005;18,301–309.
[6] Bajpai S, Gupta JP. Terror-proofing chemical process industries. Process Safe Environ Prot. 2007;85,559–565.
[7] Bajpai S, Gupta JP. Securing oil and gas infrastructure. J. Pet Sci Eng. 2007;55,174–186.
[8] Baybutt P. How can process plants improve security? Security Management. November, 2002;150–152.

[9] Baybutt P. Assessing risks from threats to process plants: Threat and vulnerability analysis. Process Saf Prog. December 2002;21(4),269–275.

[10] Baybutt P. Process security management systems: Protecting plants against threats. Chemical Engineering. 2003;110(1),48–56.

[11] Baybutt P. Making sense of cyber security. Primatech White Paper, www.primatech.com; 2003.

[12] Baybutt P. Cyber security vulnerability analysis: An asset-based approach. Process Saf Prog. December 2003;22(4),220–228.

[13] Baybutt P. Inherent security: Protecting process plants against threats. Chem Eng Prog. December 2003;35–38.

[14] Baybutt P. A scenario-based approach for industrial cyber security vulnerability analysis. Hydrocarbon Processing. March 2004;83(3),49–53.

[15] Baybutt P. Sneak path security analysis for industrial cyber security. Intech. September 2004;51(9),56–58.

[16] Baybutt P. Cyber security risk analysis for process control systems using rings of protection analysis (ROPA). Process Saf Prog. December 2004;23(4),284–291.

[17] Baybutt P. Combating cyber risk with SVA. Hydrocarbon Eng. January 2005;10(1),12–18.

[18] Baybutt P. Integrating cyber security with other sureties in a management system. ISA Technical Conference on Manufacturing & Control Systems Security; October 2005. Chicago, IL.

[19] Baybutt P. Integrating and improving cyber and physical security vulnerability analysis (SVA). First Latin American Process Safety Conference; May 27–29, 2008. Center for Chemical Process Safety. Buenos Aires, Argentina.

[20] Baybutt P. Analytical methods in process safety management and system safety engineering – Process hazards analysis. In: Haight JM. (editor) Handbook of loss prevention engineering; Wiley-VCH, Weinheim, Germany; 2013.

[21] Baybutt P. Analytical methods in process safety management and system safety engineering – Layers of protection analysis. In: Haight JM. (editor) Handbook of loss prevention engineering; Wiley-VCH, Weinheim, Germany; 2013.

[22] Baybutt P. Calibration of risk matrices for process safety. J Loss Prevent Proc Ind. 2015;38,163–168.

[23] Baybutt P. Designing risk matrices to avoid risk ranking reversal errors, Process Saf Prog. March 2016;35(1),41–46.

[24] Baybutt P. Cognitive biases in process hazard analysis. J Loss Prevent Proc Ind. September 2016;43,372–377.

[25] Baybutt P. Addressing subjectivity and uncertainty in using risk matrices. Loss Prevention Bulletin. December 2016; 252.

[26] Baybutt P. A framework for critical thinking in process safety management. Process Saf Prog. December 2016;35(4),337–340.

[27] Baybutt P. Get creative with process safety management. Chem Eng Prog. April 2017; 56–60.

[28] Baybutt P. Guidelines for designing risk matrices. Process Saf Prog. DOI 10.1002/prs.11905.

[29] Baybutt P. Guidelines for using risk matrices. Process Saf Prog. Submitted for publication; 2017.

[30] Baybutt P. Issues in security risk assessment for the process industries. J Loss Prevent Proc Ind., DOI 10.1016/j.jlp.2017.05.023.

[31] Baybutt P, Ready V. Protecting process plants: Preventing terrorist attacks and sabotage. Homeland Defense J. February 2003;2(3),1–5.

[32] Brown H. Occup Health Saf. 1998;67,172–173.

[33] Cox LA. What's wrong with risk matrices? Risk Analysis. 2008;28,497–512.

[34] CCPS. Guidelines for Chemical Process Quantitative Risk Analysis. 2nd edition, New York: Center for Chemical Process Safety/American Institute of Chemical Engineers; 2000.

[35] CCPS Guidelines for Analyzing and Managing the Security Vulnerabilities of Fixed Chemical Sites. New York: Wiley-AIChE; 2003.

[36] CCPS. Guidelines for Developing Quantitative Safety Risk Criteria. NY: Center for Chemical Process Safety/American Institute of Chemical Engineers; 2009.
[37] CIDX. Report on Cybersecurity Vulnerability Assessment Methodologies. Version 2.0. Chemical Industry Data Exchange (CIDX); November 2004.
[38] Dunbobbin BR, Medovich TJ, Murphy MC, Ramsey AL. Security vulnerability assessment in the chemical industry. Process Saf Prog. September 2004;23(3),214–220.
[39] Garcia ML. The design and evaluation of physical protection systems. Oxford: Butterworth-Heinemann; 2001.
[40] Garcia ML. Vulnerability assessment of physical protection systems. Oxford: Elsevier Butterworth-Heinemann; 2006.
[41] Guikema S D, Aven T. Assessing risk from intelligent attacks: A perspective on approaches. Reliab Eng Syst Safe. 2010;95,478–483.
[42] Haimes YY. Risk modeling, assessment, and management. Fourth edition. New York: Wiley; 2016.
[43] Hubbard DW. The failure of risk management: Why it's broken and how to fix it. Hoboken, NJ: Wiley; 2009.
[44] ISO. Risk management – Vocabulary. ISO Guide 73:2009. Geneva: International Organization for Standardization; 2009.
[45] ISO. Risk management – Principles and guidelines ISO 31000:2009. Geneva: International Organization for Standardization; 2009.
[46] Kaplan S. An introduction to TRIZ. The Russian theory of inventive problem solving. Southfield, MI: Ideation International; 1996.
[47] Kaplan, S. The words of risk analysis. Risk Analysis, 1997;17(4),407–417.
[48] Khakzad N. Reniers G. Using graph theory to analyze the vulnerability of process plants in the context of cascading effects. Reliab Eng Syst Safe. 2015:143,63–73.
[49] Lemley JR, Fthenakis VM, Moskowitz PD. Security risk analysis for chemical process facilities. Process Saf Prog. September 2003;22(3).
[50] Montibeller G, Von Winterfeldt D. Cognitive and motivational biases in decision and risk analysis. Risk Analysis. 2015; 25(7),1230–1251.
[51] Moore DT. Critical thinking and intelligence analysis (Occasional Paper). CreateSpace Independent Publishing Platform; 2013.
[52] Nolan D. Safety and security review for the process industries. 4th edition. Amsterdam: Elsevier; 2015.
[53] Reniers GLL, Dullaert W, Audenaert A, Ale BJM, Soudan K. Managing domino effect-related security of industrial areas. J. Loss Prevent Process Ind. 2008;21(3),336–343.
[54] Reniers G, Herdewel D, Wybo J-L. A threat assessment review planning (TARP) decision flowchart for complex industrial areas. J. of Loss Prevention Proc Ind. 2013;21,1662–1669.
[55] Reniers G, Sörensen K, Khan F, Amyotte P. Resilience of chemical industrial areas through attenuation-based security. Reliab Eng Syst Safe. 2014;131,94–101.
[56] Taleb NN. The black swan: The impact of the highly improbable. 2nd edition. New York: Random House; 2010.
[57] Zhuang J, Bier VM. Reasons for secrecy and deception in homeland-security resource allocation. Risk Analysis. 2010;30(12),1737–1743.

Shailendra Bajpai, J. P. Gupta

4 Security risk assessment: Some techniques

Abstract: The security risk arising due to intentional threats such as terrorism or acts of sabotage are now considered real and credible. Chemical process plants, refineries and related industries have the potential to become prime targets of terrorist organizations since they handle and process hazardous chemicals (Hazchems) under extreme operating conditions of pressure, temperature, concentration, flow rate, etc.

In this chapter, we discuss the various security risk assessment (SRA) techniques available in the literature. SRA can be carried out qualitatively by the conventional security vulnerability assessment (SVA) approach including asset characterization, threat analysis, vulnerability analysis, risk assessment, etc. Other SRAs, including security risk factor tables (SRFT) and modified SRFT models, and the rings of protection approach (ROPA) are discussed at length. It is recognized that managing security issues in chemical plants is even more challenging as the threats are dynamic in nature and their counter management strategies need to be updated from time to time. Only a limited number of researchers are working on this serious topic because it is considered that security is a state subject. However, the scientific community must think out of the box and create procedures that are fool-proof and effective. It is recognized that many of the conventional safety and security measures adopted thus far may have to be modified in light of the complex nature of the present security threats. Finally, information sharing, coordination amongst chemical process industries (CPI) operators, other allied industries, local law enforcement, health departments, etc., will be critical to manage the security risk from intentional acts.

4.1 Introduction and background

The present global situation is very difficult as violent extremism and terrorism are omnipresent. It is important to protect our critical assets, including chemical process industries, which are vital for any country. Chemical process plants, refineries and related industries have the potential to become prime targets of terrorist organizations since they handle and process hazardous chemicals (Hazchems) under extreme operating conditions of pressure, temperature, concentration, flow rate, etc. [1]. Therefore, it is essential to develop security risk assessment strategies to ensure that site security guidelines can be properly implemented.

There has been a significant increase in terrorist and other related activities directed towards chemical facilities world-wide. In January 2016, the Nigerian National Petroleum Corporation (NNPC) temporarily closed down two of its four refin-

https://doi.org/10.1515/9783110499087-004

eries, one each in central Nigeria and at Port Harcourt in southern Nigeria after militants attacked oil pipelines and caused significant supply problems [2]. In another major incident in June 2014, Iraq's largest oil refinery in Baiji was seized by ISIS, who also took foreigners as hostages [3]. The refinery was badly damaged and it will take years to have it operational again. In January 2013, armed militants attacked the Amenas gas plants in Algeria and seized hostages; this lasted over 3 days [4]. At least 39 foreign hostages were killed and many were wounded.

It is evident from the above facts that we need to protect chemical plants and their operations not only from unintentional events (accidents) but also from intentional threats, including terrorism. Chemical process safety has been given due attention, especially after the Flixborough disaster (1974) and Bhopal gas tragedy (1983). Similarly, security risk assessment of chemical plants and isolated storage has been given significant attention since the 9/11 attacks. A successful safety management program includes four very important elements [5]:
1. identification of hazards
2. prioritization of hazards
3. Consequence analysis of credible hazards
4. Emergency response planning

Similarly, in the security risk management approach, threats and vulnerabilities need to be identified and security countermeasures must be prioritized and adopted. Managing security issues in Chemical Process Industries (CPI) is even more challenging, as the threats are dynamic in nature and their counter management strategies need to be up-dated from time to time [6].

Only a limited number of researchers are working on this serious topic because security is thought of as a state subject. However, as scientists we need to think out of the box and create procedures that are foolproof and effective. In this chapter, we discuss some important security risk assessment (SRA) strategies available in the literature and point to future directions. Much work has been done on cyber and network security. However, work on SRAs related to chemical plants and related facilities is rather limited. In this chapter, the following principal SRAs will be discussed:
- classical security vulnerability assessment methodology;
- the security risk factor table (SRFT) and modified SRFT model;
- the rings of protection approach (barriers);
- miscellaneous other work.

4.2 Classical security vulnerability assessment methodology

Post 9/11 scenario, many US agencies issued security guidelines for CPIs and petroleum installations. In these guidelines a systematic security risk management approach was used to analyze security risks [7]. The process involves identifying critical assets, credible threats from different adversaries, vulnerabilities of assets and evaluating the adequacy of countermeasures. This systematic approach is often termed security vulnerability assessment (SVA). SVAs are usually performed qualitatively by a team of experts related to chemical and allied industries. The expected outcome of SVA includes identification and prioritization of risks from credible threats and selection of suitable security countermeasures. Many organizations have developed their own SVAs, which are best suited to them [8–10]. These SVAs are have largely developed in the US, but the majority of nations have yet to come up with SVAs best suited to them.

Most of these SVAs discuss threat assessment, vulnerability assessment, process hazard analysis, physical security, cyber security, gap analysis, etc. These studies are usually specific to chemical and oil installations but can be extended to other infrastructures as well. These also include roles of management in security incidents along with the emergency preparedness and its periodic assessment, etc. Physical security, cyber security and network security issues are also discussed at length in these SVAs.

It is important to mention here that most of these SVAs are voluntary initiatives by the operators and not enforced by the law. It is the responsibility of the owner/operator of a chemical plant to choose the SVA method and depth of analysis that best suits the plant's location. Differences in geographic location, target attractiveness, presence of specific type of adversaries, types of operations, and on-site quantities of Hazchems all play important roles in determining the level of SVA and the approach followed. However, there are some common elements that are independent of the SVA method used [7–10].

- Asset characterization: Characterize the facility to identify the critical assets that need to be secured, their importance and interdependencies on supporting infrastructure.
- Threat assessment: Identify and characterize threats from various adversaries and determine the risk against critical assets identified in asset characterization. Evaluate the assets in terms of attractiveness of the targets to each adversary and the related consequences.
- Vulnerability assessment: Identify potential security vulnerabilities that could threaten the asset's service or integrity. Develop plausible scenarios through which credible threats could be realized.
- Consequence analysis: Estimate the consequences in case of a successful attack both on-site or from off-site.
- Risk assessment: Based on the above steps, identify and prioritize credible security risks that need to be addressed.
- Recommendations: Identify and evaluate risk mitigation options (based on security risk assessment including cost benefit analysis) and re-assess risks to ensure that the countermeasures being applied are adequate.

Some important SVAs and related methodologies are outlined below:

- The American Chemistry Council (ACC), the Chlorine Institute, and the Synthetic Organic Chemical Manufacturers Association (SOCMA) jointly developed and

issued “Site Security Guidelines for the US Chemical Industry” in October 2001. It was further updated in 2002 [11].
- The American Institute of Chemical Engineers (AIChE)/Center for Chemical Process Safety (CCPS) published “Guidelines for Managing and Analyzing the Security Vulnerabilities of Fixed Chemical Sites” in August 2002 [12].
- The Synthetic Organic Chemical Manufacturers Association published the “SOCMA Manual on Chemical Site Security Vulnerability Analysis Methodology and Model”, in November 2002 [13].
- The American Petroleum Institute (API)/National Petrochemical and Refiner’s Association (NPRA) published “Security Vulnerability Assessment Methodology for the Petroleum and Petrochemical Industries”, in May 2003 [7]. It was further updated in 2004 and 2005.

Each of the above methodologies employs traditional risk assessment techniques and provides a well-defined, systematic framework to identify threats, security risks and vulnerabilities. Some of the important SVAs are discussed and compared in Table 4.1 [7, 10,1 1].

The American National Standards Institute (ANSI)/American Petroleum Institute (API) Standard 780 Security Risk Assessment (SRA) Methodology was published in June 2013 as a US Standard for Security Risk Assessments for Petroleum and Petrochemical Facilities [14]. The Standard represents a model for evaluating all security risks of petroleum and petrochemical infrastructure. The 2013 Standard is an update from the previous one [9] that focuses on expanding functional utility without changing the basic methodology. It can be applied to a wide range of assets, including refineries and petrochemical manufacturing plants, pipelines, and transportation facilities, etc. This standard is voluntary but has been adopted by the Kingdom of Saudi Arabia’s Ministry of Interior High Commission for Industrial Security as the mandatory security risk assessment methodology for industrial facilities in Saudi Arabia.

Moore et al. [15] developed an approach called risk analysis and management for critical asset protection (RAMCAP). It helps in defining the facilities and operations of national and regional interest for the threat of terrorism and also defines standardized methods for analyzing threats, vulnerabilities, consequences and best security practices of the industry. It basically guides the management of critical asset protection of the chemical manufacturing, petroleum refining and liquefied natural gas (LNG) sectors. It discusses two key tasks:
1. The development of a screening tool to supplement the Department of Homeland Service’s (DHS) understanding of the assets that are important to protect against terrorist attacks and to prioritize the activities.
2. The development of a standard security vulnerability analysis (SVA) framework for the analysis of threats, vulnerabilities, and consequences.

Tab. 4.1: Comparison of various SVAs

	ACC, 2001	API, 2003	CCPS, 2003
Major SVA process steps	SVA steps including project planning, facility characterization, threat assessment, vulnerability analysis and identification of countermeasures.	SVA steps including Asset characterization, threat assessment, vulnerability analysis, risk assessment and countermeasures analysis.	Five SVA steps including project planning, facility characterization, threat assessment, vulnerability analysis and identification of countermeasures. However, this method is both an asset and a scenario based approach.
Facility Characterization	Facilities and assets are broadly defined as people, information and property. Critical assets were identified and characterized by conducting site security review.	Identifying the existing layers of protection. Analysis of information, identification of hazards and their consequences and study of the technical details of facilities available and required.	Critical asset identification, hazard identification, consequence analysis, attractiveness analysis, layers of protection review and potential target list.
Threat analysis	Threat analysis was done to identify adversaries and analyze their capabilities.	Studied internal threats, external threats, internally assisted threats. Selections of threats include reasonable local, or national intelligence information.	Threat analysis and characterization was done to identify the both internal and external adversaries.
Vulnerability	A process hazard analysis is adapted to a vulnerability assessment. It includes potential causes and consequences of fires, explosion, releases and major spills of chemicals. Check list reviews were used.	Vulnerability analysis includes the relative pairing of each target asset and threat to identify potential vulnerabilities related to process security events. This involves the identification of existing countermeasures and their level of effectiveness in reducing those vulnerabilities.	It includes both asset based and scenario based approaches to identify weaknesses in the chemical facilities.
Risk assessment	Not conducted.	Security risks were evaluated qualitatively. The risk factor included consequences of a successful attack against an asset and likelihood of successful attack.	Conducted qualitatively using threat, target attractiveness, consequences and frequency as security risk factors.

Tab. 4.1: (continued) Comparison of various SVAs

	ACC, 2001	API, 2003	CCPS, 2003
Security countermeasures	This step identified the deficiencies and made recommendations based on the check list analysis and site security review. It also involved the ring of protection, cooperation with law enforcement agencies, security staff in other companies, fellow members of trade associations in order to share the threat information.	Based on the vulnerabilities identified in risk analysis, security countermeasures are identified and evaluated to prioritize for potential enhancement.	Security countermeasures were recommended after analysis of the credible risk as identified in vulnerability analysis.

Jaeger [16] developed the vulnerability assessment methodology (VAM), which is a systematic risk based approach where risk is a function of the severity of consequences of an undesired event, the attack potential, and the likelihood of adversary success in causing the undesired event. VAM consists of 13 steps: Screening, vulnerability assessment project definition, characterization of facility, derivation of consequence severity levels, threat assessment, identification of priority cases for analyses, and preparation for analyses, site survey, system effectiveness analyses, risk determination, recommendations for reduction of risk, consideration of impacts and final report preparation.

Bajpai and Gupta [1, 17] suggested that SVA can be modified to have a HAZOP type vulnerability assessment approach. They proposed the following steps for vulnerability assessment (Figure 4.1). It involves identification of the critical assets in the plant, for example, storage tanks containing Hazchems or equipment operating under extreme conditions of temperature, pressure, concentration, etc. It also includes formal threat assessment (TA) that stresses the need to identify the threats from potential adversaries in each zone, for example toxic release caused by terrorists in a tank farm area.

It suggested that all credible threats identified in TA must be considered in vulnerability assessment (VA). It implies identification of vulnerabilities within each zone and development of various possible scenarios by which the credible threats identified in TA could be realized. It then requires listing of possible consequences in the case of a successful attack and suggests that security risks be assessed by evaluating the severity of consequences and the likelihood of successful attack. They suggested that additional security measures need to be adopted in light of the nature of threats, process vulnerabilities, possible consequences and existing security measures.

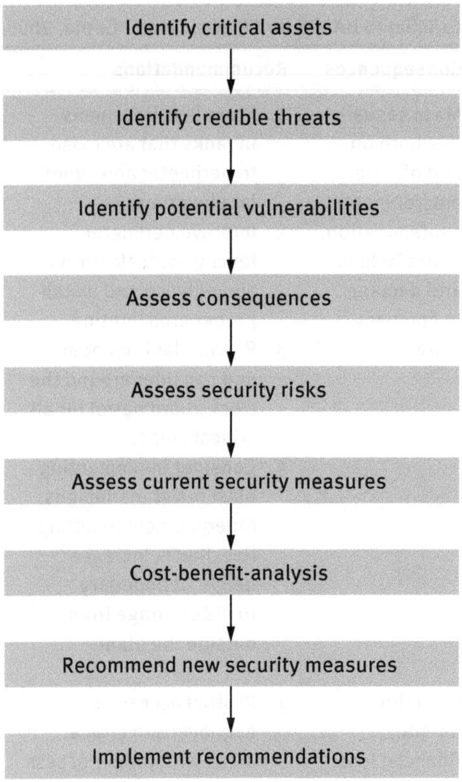

Fig. 4.1: Flowchart for SVA process [17].

Bajpai and Gupta [1, 17] concluded that a preliminary vulnerability assessment worksheet can be completed for a specific asset, detailing its threats, vulnerabilities, consequences and risk assessment (Table 4.2). However, cost-benefit-analysis and management of the change of process must be performed before new security measures are implemented.

Recently, the US President signed the law "Protecting and Securing Chemical Facilities from Terrorist Attacks Act of 2014" (CFATS Act of 2014) in December 2014 [21]. According to this law, the Department of Homeland Security (DHS) is authorized to implement the CFATS program as part of the Homeland Security Act of 2002. CPI operators must submit Security Vulnerability Assessment/Site Security Plan (SVA/SSP) to DHS through an easy-to-use online questionnaire. The operator must submit a revised SVA to DHS, in case of new threat perception in view of new chemical of interest (COI).

Tab. 4.2: Sample vulnerability assessment worksheet similar to HAZOP for CPI (Bajpai and Gupta, 2005)

Threats	Vulnerabilities	Consequences	Recommendations
Hazardous (flammable and/or toxic) substance release caused by terrorists or adversaries from outside boundaries.	1. Storage tanks are labeled and situated close to perimeter. 2. Plant is located near high population area or close to some important government establishment. 3. Plant boundary is damaged and there is no guard patrol in critical areas during daytime. 4. Projectiles could be fired from outside the boundary.	Mass casualties both on and off-site, environmental contamination, financial loss and damage to company image.	1. Store less Hazchems in tanks that are close to perimeter and avoid labeling them. 2. Improve perimeter fencing with electronic surveillance and install proper area lighting. 3. Repair plant boundary and consider around the clock guard patrol for all critical areas. 4. Consider implementing blast resistant designs for equipment handling Hazchems. Increase height of boundary to hide storage from outside the plant.
Hazardous (flammable and/or toxic) substance release caused by disgruntled employee from inside the plant.	1. Employee access to critical area is not controlled and properly managed. 2. Drain valve can be opened manually by insider. 3. Control system may be disturbed/damaged. 4. Poor labor relations in the plant. 5. No policy to conduct background checks on employees.	Casualties on-site, financial loss, environmental contamination, loss of confidence in employees and damage to company image.	1. Restrict access of employees to critical areas with proper access control procedure in place. 2. Consider installing valve locks. 3. Restrict access to control system with password control. Improve physical security near control system 4. Maintain good labor relations in the plant. Address genuine demands of employees in timely manner. 5. Conduct background checks on employees and share the information with local law enforcement agencies.

The revised SVA needs to be completed, if the new COI is located near or within critical assets and thereby increases the overall security risk of the facility. The DHS reviews the SVA/SSP to determine whether it satisfies all applicable risk-based performance standards (RBPS). The chemical operator has to indicate the various factors of chemical security assessment tool (CSAT) and the use of COI.

In the vulnerability assessment section of CSAT the operator needs to answer the questions related to detect, delay, response and cyber security measures, policies and procedures for identified vulnerabilities. The detection security measures include closed circuit television (CCTV), intrusion detection systems (IDS), proper lighting, process controls and alarms, inventory control, personnel or protective force, etc. The delay security measures include fencing of walls, vehicle barriers, locking mechanism, access control, screening and inspections, etc.

The response measures for identified vulnerabilities must address information related to the emergency response plan (ERP) within their facility, prepared to respond to security an incident that can be handled within the local law enforcement agency and first responders. The ERP may include crisis, management plans, specific threat procedures, communication equipment and outreach with local law enforcement, etc.

The CSAT tool also details a section on cyber security measures for identified vulnerabilities. The cyber security measures include policy and procedure, access control, password management, personnel security training and awareness, cyber security controls, disaster recovery, etc. Similarly, in the last section operators are asked to provide measures related to policies, procedures, and resources and identified vulnerabilities. These may include inspection testing, maintenance programs, security awareness and training programs, incident reporting and investigations, record keeping, etc.

Lee et al [19] discussed the risk assessment program against terrorism in the Republic of Korea. This study focused on assessing the security risks at chemical facilities and public utilities. It revealed that most chemical facilities remain unsecured despite their clear vulnerability to terrorism in Korea. They stressed the need to implement effective chemical terrorism response plans to enhance security. This study modified conventional methods for assessing the risk against terrorism. This work introduced chemical terrorism response technology, a prevention plan, and new countermeasures to mitigate by using suggested risk and vulnerability assessment method. This work is a modified SVA suited to the Republic of Korea.

4.3 The security risk factor table (SRFT) model

Bajpai and Gupta [1, 17] showed that the security risk status of a chemical plant can also be assessed by developing a security risk factor table (SRFT). This model helps assess the current security risk status of a plant and can be used as a pre-screening tool before initiating time consuming formal SVAs and other SRAs. In the SRFT model, important risk bearing parameters such as location, visibility, ownership, presence of Hazchems, etc. (Table 4.3) are considered and rated on a scale from 0 to 5, with 0 being the "lowest risk" and 5 the "extreme". The grading can be qualitatively completed by the experts who, based on their experience, can assign the score to a given risk parameter as presented in Table 4.4. The total risk score obtained from SRFT helps assess the current security risk status of the facility (as per Table 4.4).

Tab. 4.3: Security risk factor table (Bajpai and Gupta, 2005)

Risk factors	Range of security points				Actual points
Location	Rural	Urban		High density	
	1	2,3,4		5	
Visibility	Not visible	Low	Medium	High	
	0	1,2	3,4	5	
Inventory	Low	Medium	Large	Very large	
	1	2	3,4	5	
Ownership	Private	Public/Co-operative		Government	
	1	2,3		4,5	
Presence of chemicals which can be used as precursors for WMD	Absence		Presence		
	0		5		
Worst case impact on-site	Negligible	Low	Moderate	Severe	
	0	1	2,3,4	5	
Worst case impact off-site	Negligible	Low	Moderate	Severe	
	0	1	2,3,4	5	
History of security incidents	Nil	Few		Frequent	
	0	1,2,3		4,5	
Presence of terrorist groups	Absence	Few		Large no.	
in region	0	1,2,3		4,5	
Existing security measures:	High level	Ordinary		Poor/none	
– Access control	1	2,3		4,5	
– Perimeter protection	1	2,3		4,5	
– Mitigation potential	1	2,3		4,5	
– Proper lighting (all over)	1	2,3		4,5	
– Use of metal detector/x-ray/	1	2,3		4,5	
CCTV	1	2,3		4,5	
(at entrance and at all critical locations)					
Personal preparedness	Well prepared	Average		Poor	
and training	1	2,3		4,5	
				Total score	

Tab. 4.4: Security risk rankings (Bajpai and Gupta, 2005) (based SRFT score)

Current security risk status	Actual points obtained	Recommendations
Low	<15	Maintain security awareness without excessive concern.
Moderate	16–30	Review and update existing security procedures in light of possible threats.
High	31–45	Identify risk drivers that can be reduced with reasonable controls. Conduct threat and vulnerability analysis and work with law enforcement agencies to enhance security.
Extreme	>45	Initiate aggressive risk-reduction activity, in consultation with law enforcement agencies. Conduct threat and vulnerability analysis.

Bajpai et al. [20] further modified the SRFT model by applying the concepts of fuzzy logic. They explained that in the SRFT model experts give a crisp integer score to all risk parameters. Therefore, actual scores obtained are prone to the human subjectivity involved in making the decision. In other words, the score assigned to a risk parameter may vary from one expert to another. Therefore, in order to reduce human subjectivity, fuzzy set theory was applied in the SRFT model. Fuzzy scores were used in place of the crisp integer scores of the SRFT model. All the risk parameters of the SRFT model were first fuzzified and later defuzzified to obtain the risk score.

Tab. 4.5: Modified security risk factor table (Bajpai and Gupta, 2010)

Risk Factors	Range of Security Points				Expert score	Defuzzified Score
Location	Rural (0,0,1,2)	Urban (1,2,3,4)	High density (3,4,5,5)			
Visibility	Not visible (0,0,1,2)	Low (0.5,1.5,2.5,3.5)	Medium (2,3,4,5)	High (3.5, 4.5,5,5)		
Inventory	Low (0,0,1,2)	Medium (0.5,1.5,2.5,3.5)	Large (2,3,4,5)	Very large (3.5, 4.5,5,5)		
Ownership	Private (0,0,1,2)	Public/co-operative (1,2,3,4)	Government (3,4,5,5)			
Presence of chemicals that can be used for inflicting heavy casualties	Low Quantity (0,0,1,2)	Medium Quantity (1,2,3,4)	Large Quantity (3,4,5,5)			
Worst case impact on-site	Negligible (0,0,1,2)	Low (0.5,1.5,2.5,3.5)	Moderate (2,3,4,5)	Severe (3.5, 4.5,5,5)		

Tab. 4.5: (continued) Modified security risk factor table (Bajpai and Gupta, 2010)

Risk Factors	Range of Security Points				Expert score	Defuzz-ified Score
Worst case impact off-site	Negligible (0,0,1,2)	Low (0.5,1.5,2.5,3.5)	Moderate (2,3,4,5)	Severe (3.5, 4.5,5,5)		
History of security incidents	Nil (0,0,1,2)	Few (1,2,3,4)	Frequent (3,4,5,5)			
Presence of terrorist groups in region	Absence (0,0,1,2)	Few (1,2,3,4)	Large no. (3,4,5,5)			
Existing security measures:	High level	Ordinary	Poor / None			
– Access control	(0,0,1,2)	(1,2,3,4)	(3,4,5,5)			
– Perimeter protection	(0,0,1,2)	(1,2,3,4)	(3,4,5,5)			
– Mitigation potential	(0,0,1,2)	(1,2,3,4)	(3,4,5,5)			
– Proper lighting (all over)	(0,0,1,2)	(1,2,3,4)	(3,4,5,5)			
– Use of Metal detector/ x-ray/CCTV (at entrance and at all critical locations)	(0,0,1,2)	(1,2,3,4)	(3,4,5,5)			
Personal preparedness and training	Well prepared (0,0,1,2)	Average (1,2,3,4)	Poor (3,4,5,5)			
				Total score =		

In the modified SRFT model, two linguistic scales were used. Trapezoidal fuzzy numbers were used for 3 points: low (0, 0, 1, 2), medium (1, 2, 3, 4) and high (3, 4, 5, 5). The four-point scale used was: low (0, 0, 1, 2), medium (0.5, 1.5, 2.5, 3.5), moderately high (2, 3, 4, 5) and high (3.5, 4.5, 5, 5). The modified SRFT model is presented in Table 4.5. The evaluation for the actual security score for each risk parameter as given by the experts was performed for the scales mentioned above. A test case of a refinery was also used to show the application of the modified SRFT model in SRA. In this study, all risk parameters were given the same weightages.

4.4 Rings of protection approach (barriers)

ROPA (Figure 4.2) is used in SRA very similarly to how layers of protection analysis (LOPA) is. In LOPA, various layers of protection are added to the process in order to lower the frequency of undesired event. Similarly, in ROPA, rings of protection are added to enhance the security of critical assets such that the likelihood of a successful attack is reduced. The objective of ROPA is to provide sufficient layers of protection to a critical asset such as private security, access control, physical security, network security, etc., to an undesired event.

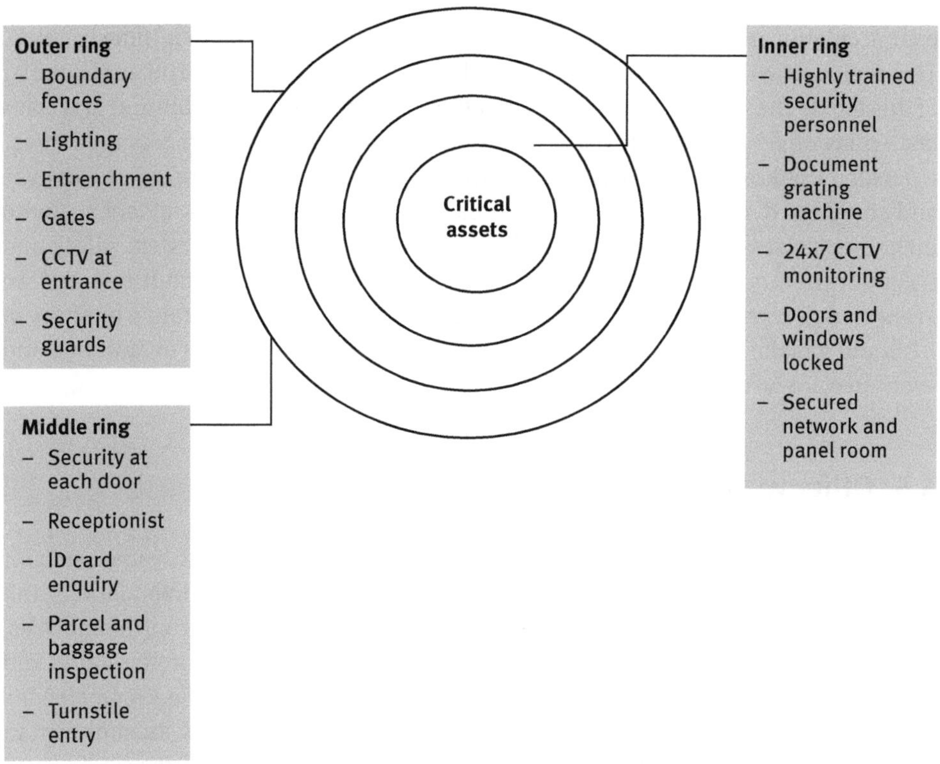

Fig. 4.2: Rings of protection (adapted from ACC, 2001) [8].

Reniers et al. [21] described a security risk assessment and methodology that was developed for use in the chemical and process industries in Belgium. The method employs a risk based approach according to design principles for object oriented protection. The approach is beneficial for workers in the chemical industry because of the familiarity with safety models and concepts in this particular industry. The model combines ROPA [22] with generic security practices including management and procedures, security technology (e. g., CCTV, fences and access control), and human interactions (proactive as well as reactive). They suggested that chemical plant operators wishing to assess their security situation should initially perform a gap analysis between the actual state of the plant and the ideal security situation. This will help them in assessing the existing security situation in the plant.

Kalantarniya et al. [23] attempted to demonstrate the testing and modeling of the BP Texas City refinery accident using dynamic risk assessment methodology as a predictive tool for accident occurrence. Dynamic risk assessment is a novel approach which integrates a Bayesian failure updating mechanism with the consequence assessment. The implementation of this methodology to the BP Texas City refinery incident proves that the approach has the ability to learn from near miss incidents, past accidents and to predict event occurrence likelihood in the next time interval. This tool is heavily dependent on the incident and near miss data. The applicability of this tool to the BP case study shows that the accident was predictable and may have been prevented had dynamic risk assessment been applied to the process industries.

Later, Khakzad et al. [24] came up with a bow tie (BT) model approach to analyze and study the dynamic risk. The BT approach represents the total accident scenario including causes and consequences. Ferdous et al. [25] analyzed system safety and risks uncertainty using a BT diagram. A BT diagram combines a fault tree and an event tree to represent the risk control parameters. However, qualitative analysis of BT is still a major challenge as it follows the traditional assumptions of fault tree and event tree analyses.

4.5 Other works on security risk management

Leitch et al [25] discussed the identification and application of security measures for petrochemical industrial control systems. This paper presents recommendations and insights from over 100 security risk assessment (SRA) and process control analyses, using requirement baselines extracted from the National Institute of Standards and Technology (NIST) special publication 800–53 (and Appendix A), the *Recommended Security Controls for Federal Information Systems and Organizations*, in conjunction with NIST special publication 800–82, *Guide to Industrial Control Systems(ICS) Security*, to provide the bridge in application of 800–53 controls to IC/SCADA. The paper identifies how current and projected malevolent threats posed by insiders, outsiders, collusion, and system induced threats can erode system performance in terms of shut downs, sabotage, production disruption and contamination.

Moore [26] explained the importance of a new risk management paradigm for chemical plant security that would require a different form of analysis than accidental risk assessment methods. He reviewed the concepts of SVA and security management principle. He explained that deliberate release risk could be managed by many of the same or similar strategies to accidental release risk. According to Moore, traditional security countermeasures, such as physical security features and cyber security measures must integrate with safety strategies to result in a single process risk management strategy.

Whiteley et al. [27] studied the level of safety provided by existing plant equipment and safety systems in response to a terrorist attack. They suggested considering the terrorist or criminal threat from a process, rather than security, point of view. They explained that all process plants are designed to deal with unintentional events such as equipment failure; loss of utilities, fire exposure from spills, etc., which threaten safe operation of the facility. However, existing safety systems were not designed to address acts of sabotage or a thinking adversary. They suggested integrating the results of SVA and HAZOP to automatically produce threat scenarios. They emphasized the utility of mapping inventories of Hazchems in the process in terms of explosive energy or fire. They concluded that work is required to determine how existing process hazard analysis (PHA) methods could be utilized to address deliberate threats. They also suggested determining what changes in equipment, policy and procedures could be implemented to minimize the impact of a terrorist attack.

In India, the National Disaster Management Authority (NDMA), of the Government of India, issued the guidelines for the Management of Chemical (Terrorism) Disasters in 2008 [28]. It is a detailed document that illustrates the historical background of chemical terrorism and chemical warfare agents being used in World Wars I and II. This guideline mainly focuses on disaster prevention measures such as surveillance, intelligence, relief, reconstruction, recovery, disaster management cycle and preparedness. The guidelines also cover the various aspects of storage, transportation, safety and risk reduction of chemicals. It also covers the mechanism for the management of chemical disasters. The desired outcome of the guidelines is to reduce the number of deaths due to chemical terrorism.

This guideline pointed out the gaps in the existing rules and regulatory system of chemical disasters and explained the need to enhance and monitor existing management policy. This includes risk and vulnerability assessment, data collection, mechanisms for surveillance, environmental health documentation, etc. The document also discussed the gaps that exist in the form of education, availability of personal protective equipment, decontamination, communication and networking. In order to fill the gaps in disaster management, the guidelines suggest frequent training programs for awareness and information for various stake holders and decision makers. It contains the response protocols to communicate with the public through media management. This protocol can also be used in the event of chemical disaster victims for immediate decontamination.

It also illustrates the management of chemical disasters through basic steps such as prevention, mitigation and management of illness and injury caused by chemical agents used in terrorist activities. Potential harmful effects of various chemicals (like panic reaction, chemical burns, injuries, physical and social trauma and effects to the environment) are described in the guidelines. They include chemical warfare agents, dual use chemicals, toxic industrial chemicals, hazardous chemicals, agriculture chemicals and other poisonous chemicals. The documentation of post disaster events was also developed and detailed in the guidelines. How to strengthen public–private participation in securing chemicals, provide insurance cover, management of hazardous waste, and mitigation is also mentioned.

The guidelines also discuss various aspects of response, rehabilitation and recovery and post documentation. The response section explains alert/notification of disasters, evaluation of the scale of disaster, possible emergency response from various departments. This may include transport, fire service, communication and medical emergency departments, etc. The rehabilitation and recovery section explains the activities that need to be followed after disasters, such as health needs, casualty management, decontamination, reconstruction, environment measurement, etc. The guidelines strongly emphasize the need for constant upgrading, which is mandatory in disaster management.

4.6 Conclusions and future work

The security risks arising due to intentional threats are viewed seriously in the chemical process industries. There have been several incidents targeting chemical facilities by extremists, militants and internal adversaries worldwide.

It is important to develop an effective SRA methodology that identifies and prioritizes security hazards and associated potential vulnerabilities. There are several SVAs available in the literature that are largely qualitative in nature and address security risk management strategies. These SVAs have been extensively discussed and compared in this chapter. It is important that SVAs are updated regularly for new threat perception. Recently, RAMCAP has become standard for petrochemical facilities in the US and the same has also been adopted by Saudi Arabia. In 2014, the US passed a law entitled the "Protecting and Securing Chemical Facilities from Terrorist Attacks Act of 2014" (CFATS act of 2014).

In India, National Disaster Management Authority (NDMA), of the Government of India, also issued guidelines for the management of chemical (terrorism) disasters in 2008. It is important to mention here that effective SRA depends on how well the threats are understood and information shared amongst important functionaries. Plausible scenarios must be developed and prioritized for improved security action.

It is important to develop a security data base for CPI. In the case of any incidents, the information should be shared amongst chemical facility operators globally and quickly through online software. The SRFT model and the modified SRFT model will also help in the identification and prioritization of security risks. These models can be expanded to different industries, processes, geographic locations, etc.

It is also important to implement existing safety tools that can be used for security purpose as well. An SVA worksheet can be developed for different parts of plants for different threats and vulnerability, very similarly to what is used in hazard and operability studies (HAZOP). Fault tree analysis (FTA) and event tree analysis (ETA) can also be developed to enhance security countermeasures in terms of detect, delay, response and mitigation. Finally, information sharing, coordination amongst CPI operators, other allied industries, local law enforcement, health department, etc., will be critical to manage security risks from intentional acts.

Acknowledgment

The authors would like to thank Mr. Abhishek Verma, Mr. R. R. Marlar and Mr. S. S. Rao for their overall assistance during the preparation of this chapter.

References

[1] Bajpai S, Gupta JP. Site security for chemical process industries. J Loss Prev Process Ind. 2005;18:301–309.

[2] BBC News. Nigeria to break up loss-making state oil firm NNPC, 2016. Accessed in Nov 2016, at http://www.bbc.com/news/world-africa-35721693.

[3] Daily Mail. ISIS “seizes Iraq’s largest oil refinery” and kidnaps 100 foreigners but country’s PM insists “we have regained the initiative and are striking back,” 2014. Accessed in Nov, 2016 at http://www.dailymail.co.uk/news/article-2661134/Iraqi-PM-says-government-forces-striking-ISIS-plans-wave-terror-attacks-Baghdad.html.

[4] Statoil News. Publication of the investigation report on the Amenas terrorist attack, 2013. Accessed in Nov 2016 at http://www.statoil.com/en/NewsAndMedia/News/2013/Pages/12Sep_InAmenas_report.aspx.

[5] Crowl D, Louvar JF. Chemical process safety: Fundamentals with applications. New Delhi: Pearson, 2014.

[6] Bajpai S, Gupta JP. Protecting chemical plants from terrorist attacks. Chemical Weekly. 2005;L(34):209–213.

[7] American Petroleum Institute. Security Guidelines for the Petroleum Industry. 2003, Washington DC. Accessed in Nov 2016 at http://www.api-ec.api.org/filelibrary/Security_Guidance2003.pdf.

[8] American Chemistry Council, Chlorine Institute, and Synthetic Organic Chemical Manufacturers Association. Site Security Guidelines for the US Chemical Industry. 2001, Washington DC. Accessed in Nov 2016 at http://www.accnewsmedia.com/docs/100/89.pdf.

[9] American Petroleum Institute. National Petrochemical & Refiners Association. Security vulnerability assessment methodology for the petroleum and petrochemical industries. Washington DC, 2004.

[10] Centre for Chemical Process Safety, American Institute of Chemical Engineers. Guidelines for Analysing and Managing the Security Vulnerabilities of Fixed Chemical Sites. New York, 2003, 10016–5991.

[11] American Chemistry Council. Implementation guide for responsible care security code of management practices site security and verification, 2002. Accessed in November 2016 at

http://www.nj.gov/dep/enforcement/security/downloads/ACC%20Responsible%20Care%20Site%20Security%20Guidance.pdf.
[12] Centre for Chemical Process Safety, American Institute of Chemical Engineers. Guidelines for Managing and Analyzing the Security Vulnerabilities of Fixed Chemical Sites. New York, 2002.
[13] Synthetic Organic Chemical Manufacturers Association, Inc. (SOCMA). Manual on Chemical Site Security Vulnerability Analysis Methodology and Model, 2002. Accessed in Nov, 2016 at http://cchealth.org/hazmat/pdf/calarp/appendix_i.pdf.
[14] Moore DA. Security risk assessment methodology for the petroleum and petrochemical industries. J Loss Prevent Proc Ind. 2013; 26:1685–1689.
[15] Moore DA, Fuller B, Hazzan M, Jones J. Development of a security vulnerability assessment process for the RAMCAP chemical sector. J Hazardous Materials. 2007;142:689–694.
[16] Jaeger CD. Chemical facility vulnerability assessment project. J Hazardous Materials. 2003;104(1–3):207–213.
[17] Bajpai S, Gupta JP. Securing oil and gas infrastructure. J Pet Sci Eng. 2007:55(1):174–186.
[18] Homeland Security. Chemical Security Assessment Tool (CSAT) 2.0. Security Vulnerability Assessment/Site Security Plan Instructions, 2016. Accessed in December 2016 at https://www.dhs.gov/csat-sva-ssp.
[19] Lee Y, Kim J, Moon I. Development of a risk assessment program against terrorism in Republic Korea. ICheme symposium series no. 153, 2007.
[20] Bajpai S, Sachdeva A, Gupta JP. Security risk assessment: Applying the concepts of fuzzy logic. J Hazardous Materials. 2010;173:258–264.
[21] Reniers GLL. Multi-plant safety and security management in the chemical and process industries. New York: Wiley VCH, 2010.
[22] Reniers G, Lerberghe P V, Guljk CV. Security risk assessment and protection in the chemical and process industry. Process Saf Prog. 2014;34:72 –83.
[23] Kalantarnia M, Khan FI, Hawboldt K. Modelling of BP Texas City refinery accident using dynamic risk assessment approach. Process Saf Environ Protect- 2010;88(3):191–199.
[24] Khakzad, Khan N, Amyolle P, Cozzani V. Risk management of domino effects considering dynamic consequences analysis. Risk Analysis. 2014;34(6):1128–1138.
[25] Ferdous R, Khan F, Sadiq R, Amyotte P, Veitch B. Analyzing system safety and risks under uncertainty using a bow-tie diagram: An innovative approach. Process Saf Environ Protect. 2013;91:1–18.
[26] Leith HM, Piper JW. Identification and application of security measures for petrochemical industrial control systems. J Loss Prevent Process Ind. 2013;26:982–993.
[27] Whiteley JRR, Mannan SM. Initial perspectives on process threat management. J Hazardous Materials. 2004;115:163–167.
[28] National Disaster Management Authority Government of India. Management of Chemical (Terrorism) Disaster. New Delhi, 2009.

Gabriele Landucci, Francesca Argenti, Valerio Cozzani

5 A methodology for the evaluation of attractiveness with respect to external acts of interference dedicated to the chemical and process industry

Abstract: Among the most critical targets of malicious acts of interference, process plants play a relevant role, due to the high quantities of dangerous substances that may be present on these sites. External attacks on this type of facility may lead to severe consequences, both considering fatalities among the population and socio-economic fallouts. The concern about the possibility of such events has increased greatly in the last years; besides the inherent attractiveness of the industrial facility, the aversion of the local population towards the plant location may be a key factor enhancing the likelihood of an external attack. The present study is aimed at evaluating the "attractiveness" of the plant through a structured approach that takes into account i) the potential consequences of accidental scenarios that may be triggered by malicious acts of interference and ii) a set of relevant factors that influence the targeting logic on the basis of strategic and geopolitical considerations. The procedure for attractiveness assessment was exemplified through the analysis of case studies, which demonstrated the importance of not limiting plant attractiveness assessment to a consequence based evaluation, stressing the importance of geopolitical, social and economic incentives.

5.1 Introduction

In the past, chemical process facilities were believed to be extremely unlikely targets of terroristic acts when compared to public malls, railway stations, and other crowded locations. After the New York City attacks of "9/11", the security of sites where relevant quantities of hazardous chemicals are stored or processed became a concern [1]. In fact, the hazard posed by security threats to this type of facility, in terms of disruption of operations, destruction of property, injury or loss of life are somehow comparable to those coming from major accidents due to internal causes [2]. To give an example, major accidents may be triggered by external attacks carried out using military explosives or improvised explosive devices. Therefore, security risks started to be included in formal risk assessment [3]. The attacks perpetrated in France in 2015 against the production site of a chemical company [4] confirmed the credibility of terroristic threat to industrial facilities located in western countries.

In USA, security-related legislation was enacted. The Department of Homeland Security (DHS) is required to analyze vulnerabilities and establish risk based security

https://doi.org/10.1515/9783110499087-005

performance standards for critical infrastructure and facilities; while facility owners and operators are required to prepare a security vulnerability assessment (SVA) and a facility security plan, according to the prescriptions of the Chemical Facility Anti-Terrorism Standards (CFATS) [5].

In Europe, the "European Programme for Critical Infrastructure Protection (EPCIP)" [6] promotes the prevention, preparedness and response to terrorist attacks involving installations of the energy (electricity, oil and gas) and the transport (road, rail, air, inland waterways and ocean and short-sea shipping and ports) sectors. On the other hand, European Seveso-III Directive [7] concerning major accident hazards focuses on safety-related issues and does not address the need for a security analysis or for security countermeasures in industrial installations that may be considered attractive or vulnerable targets of terrorist attacks. Hence, no detailed guidelines are yet available for the security of chemical and process plants in the EU.

Relevant research efforts were devoted to the development of methods for security risk assessment dedicated to the process industry. Early work on the topic started after 9/11, with the development of the so-called security vulnerability assessment methodologies by professional organizations [8, 9], governmental institutes [10, 11], consultancy firms and multinational companies, which were qualitative in nature. In parallel, a number of scholars developed or promoted simplified methods to assess the level of security risk faced by industrial facilities of the chemical and O&G sector [3, 12–14], while some others focused on the psychology of individual terrorists or group processes [15], considering also arguments rooted in economic and socio-psychological dimensions of human motivations. Few semi-quantitative methodologies have been proposed or adopted in practice for the security risk assessment (SRA) of different types of facilities. Methods developed by the American Petroleum Institute [16], the American Institute of Chemical Engineering [9] and Sandia National Laboratories [17] are today among the more frequently applied.

Factors typically accounted for in conventional SRA methods include the threat, the attractiveness of the asset to adversaries, the possible consequences and impacts of an incident and the degree of vulnerability [3, 16, 18]. In particular, SRA methodologies are aimed at the characterization of target facility, threat agent, threat (or attack mode) to support the impact estimation of potential attacks. Therefore, a deeper investigation into the attractiveness of process facilities, determining whether security triggered scenarios are credible, is still lacking. Literature methods do not systematically address attractiveness of process facilities accounting for the specificity of hazards. Moreover, the integration of this type of evaluation with aspects related to the social, economic and political contest is not yet available and would be beneficial for a more systematic evaluation of process facilities attractiveness.

In the present contribution, the attractiveness of industrial facilities as potential targets is proposed as a proxy to the likelihood of attack in order to support a security risk screening analysis. A novel formulation of a previously developed method [19, 20] is herein presented to support the semiquantitative attractiveness assessment. Besides technical considerations, nontechnical factors are accounted for and weighted to more adequately depict the motives and triggers that play a role in determining attractiveness.

5.2 Methodology for attractiveness assessment in chemical facilities

5.2.1 Overview of the methodology

The present contribution illustrates a semiquantitative methodology to assess the attractiveness of process facilities to malevolent external attack. The method was developed in order to have input data easy to gather, which could be derived from documents available to plant operators, in order to facilitate method application and to obtain a quick but exhaustive screening tool. According to [9, 16], the evaluation of attractiveness is a critical part of SRA aimed at:

i. supporting the prioritization of resources
ii. identifying possible security criticalities for a given facility

According to American Petroleum Institute (API) 780 standard [16], assessing attractiveness means estimating the value of a target to a specific source of threat. In agreement with this definition, the proposed assessment was centered on the target characteristics, determining its potential to lead to casualties and property damage, considered as the objective basis to quantitatively estimate its "value". Consideration was then given to other less objective aspects contributing to increasing the target value for symbolic or strategic reasons. Hence, the evaluation procedure applies both quantitative and qualitative techniques to provide a relative ranking of the relevant aspects associated with plant attractiveness.

The methodological framework adopted in the present study is summarized in Fig. 5.1.

As shown in Fig. 5.1, the attractiveness assessment proposed in this study is based on the evaluation of an overall attractiveness index (I_A), which, in turn, depends on two different contributions: the "hazard based" facility attractiveness index (I_H) and a site-specific "induction index" (ζ). The introduction of indices is aimed at defining parameters for supporting the ranking of specific features of a given facility (e. g., inherent hazard, territorial vulnerability, attractiveness).

The I_H index is an indicator of the quantifiable value of an installation as target, in terms of major accidents and damage potential in case of external attacks. Hence,

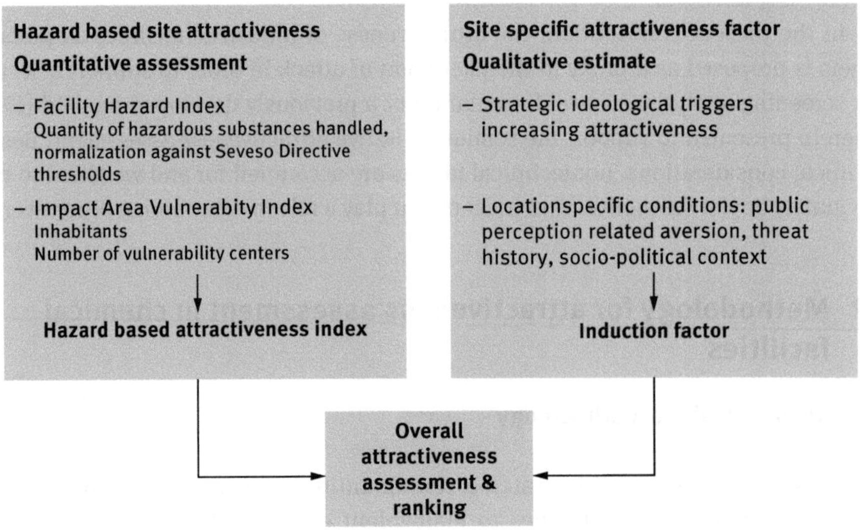

Fig. 5.1: Methodological approach for the evaluation of overall facility attractiveness

I_H is related to the magnitude of the potential consequence of an attack to the target. The quantitative evaluation of I_H is performed accounting for both the process facility inherent hazard, based on the analysis of the hazardous material inventories, and the vulnerability of the area surrounding the facility under analysis that might be impacted by an accident triggered by an external attack. More details are reported in Section 5.2.2.

The induction index ζ is then introduced to modify the index I_H, accounting for specific issues that may affect the final value of the index. The aspects considered in the evaluation of ζ are socio-economic and cultural factors, thus it is associated with nontechnical triggers. The index ζ is calculated as a function of the "overall attractiveness increase index" F as follows:

$$\zeta = 1 + F \tag{1}$$

The calculation of the F index is based on the systematic identification and scoring of a set of threat triggers and deterrence factors, as explained in Section 5.2.3. The metric for the F index was set in order to obtain safe side evaluations and it is only considered as a worsening element. It is worth remarking that the presence of security countermeasures and preventive barriers (which may decrease *F* in this perspective) is not taken into account at this stage of the work.

The indexes obtained through the application of the presented methodology are based on the assignment of scores, following the guidelines described in Sections 5.2.2 and 5.2.3

Once the indexes I_H and ζ are determined, following the calculation guidelines provided in Sections 5.2.2 and 5.2.3, the overall attractiveness index I_A is evaluated as follows:

$$I_A = I_H \times \zeta \tag{2}$$

In order to rank the overall attractiveness index, I_A, three levels of attractiveness are defined: high, medium and low, on the basis of the overall index value. Table 5.1 reports the correspondence between the evaluated scores for I_H and I_A and the qualitative ranking levels.

Tab. 5.1: Qualitative ranking associated to hazard based attractiveness index I_H and overall attractiveness index I_A

Index	Score Range	Qualitative ranking
I_H	2–5	Low
	5–8	Medium
	> 8	High
I_A	2–5	Low
	5–8	Medium
	> 8	High

The methodology outline above might be useful to rank facilities in the same geographical area or identify critical industrial sectors. Moreover, criteria and priorities for emergency planning and for protection actions might be derived from the result of the proposed analysis.

5.2.2 Hazard based attractiveness index

As stated in Section 5.2.1, the I_H index is meant to describe in a sound way the value of the installation in terms of major accidents and severe damage potential. The quantitative evaluation of I_H is performed according to the scheme summarized in Fig. 5.2, which represents a further development and re-arrangement of the method introduced in a previous study [19, 21].

As shown in Fig. 5.2, I_H is function of two main subindexes, namely I_{FH}, the process facility hazard index, and I_{TV}, the index that accounts for vulnerability of territory surrounding the site:

- I_{FH} is based on the analysis of the hazardous material inventories of the facility under examination, which are related to potential damage connected with the facility. A normalized substance (or substance category) hazard index is employed to characterize the inherent damage level of a facility due to the stored and processed quantities of hazardous materials (see Fig. 5.2);

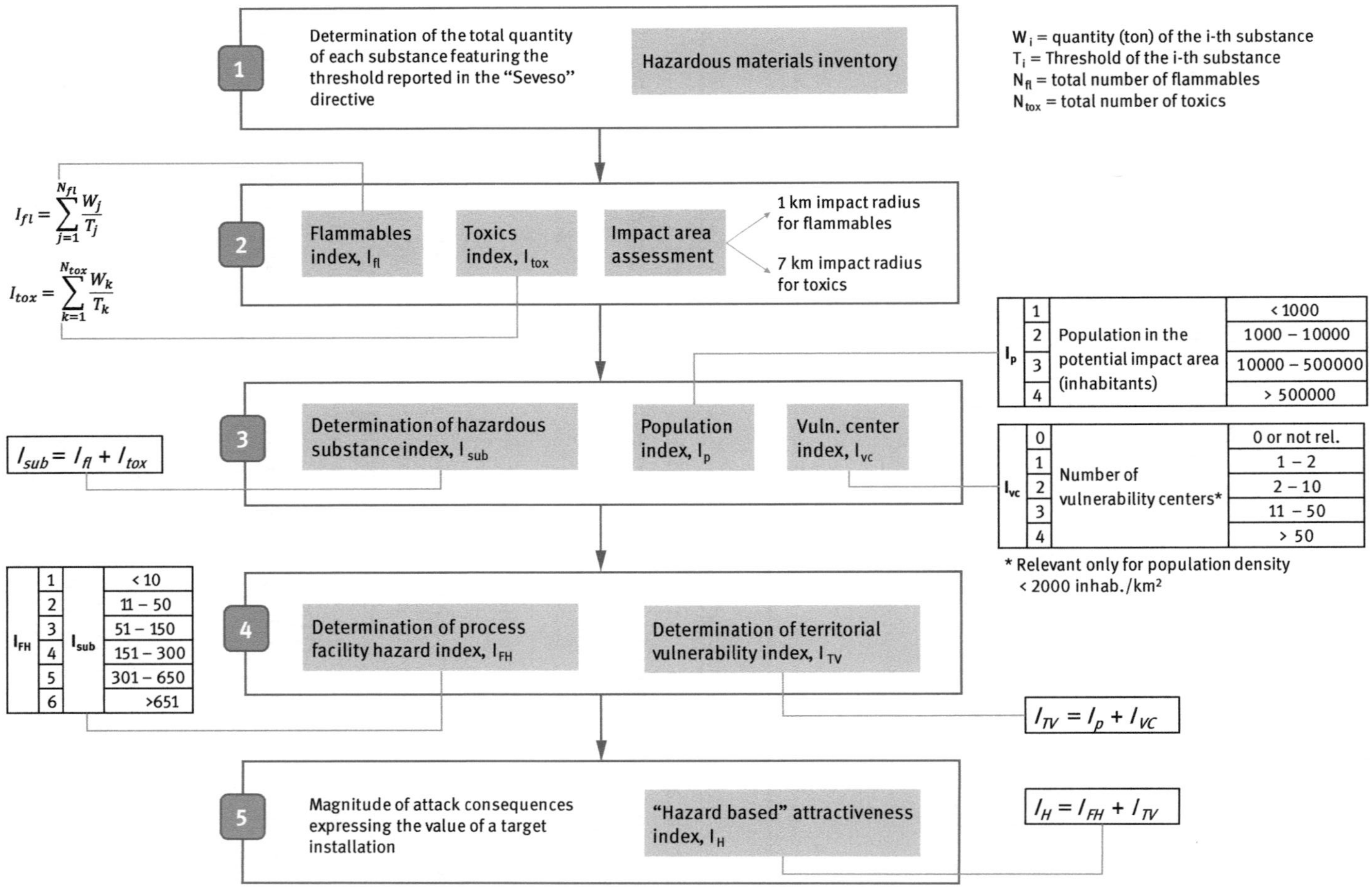

Fig. 5.2: Summary of the procedure for the hazard based facility attractiveness assessment

- I_{TV} is related to the vulnerability of the area surrounding the facility under analysis that might be impacted by an accident triggered by an external attack. For a preliminary assessment, the impact area can be approximated on the basis of the plant substance inventory (see Fig. 5.2). The vulnerability of the area around the plant is then related not only to the population density, but also to the possible presence of vulnerability centers (i. e., sites in which the population density is much higher than in normal residential areas, such as hospitals, schools, malls, etc.).

More details on the methodology for the quantitative evaluation of I_H are reported elsewhere [19, 21].

5.2.3 Estimation of nontechnical triggers: induction index

The hazard based attractiveness index presented in Section 5.2.2 provides a metric for the attractiveness based on the destructive potential of a successful attack. However, threat agents may have many other incentives to attack a facility, hence also nontechnical aspects and triggers need to be taken into account [22, 23].

An exhaustive threat characterization in the area under analysis would investigate the threat history, intent, motivation, and capabilities, detailing the known patterns of potential adversaries, in order to provide a sound basis for the identification of targeting logics and preferences for different types of threat agents. However, it is worth mentioning that, according to API [16], in most cases the best available information is generic and nonspecific with respect to the facility or to the location. Thus, a detailed threat assessment trough characterization of adversaries' classes, ideologies, political goals and primary target audience is beyond the scope of the present study.

In order to evaluate the general influences that may play a role in increasing the perceived value of a potential target facility, a hierarchical approach was followed in order to determine the overall attractiveness increase index F, which is adopted in Eq. 1 to calculate the induction index (ζ).

Table 5.2 reports the different criteria accounted for in the evaluation of F. The selection of the most relevant criteria was made through a screening of relevant literature studies (see references cited in Tab. 5.2). For example, in the literature it is often emphasized that targets often tend to be chosen for their symbolic value rather than their absolute value or utility (in terms of maximizing casualties per se) [24]. On the other hand, although seeking for maximum intimidation effect, terrorists, especially political-religious groups who seek support from the broader population, avoid attacks that may negatively affect outside stakeholders and supporters, thus risking a backlash [23]. In the definition of each item listed in Table 5.2, a general perspective is kept to deal with adversaries' incentives or deterrents to perpetrate an attack, making considerations valid for different categories of threat agents: political or religious terrorists, disgruntled insiders or contractors and activists.

It is, however, worth remarking that the problem complexity may not exclude a cross influence among aspects considered in the present study, which was herein neglected for the sake of method simplicity.

A uniform scoring system is adopted to consider the influence of the aspects summarized in Tab. 5.2. The overall attractiveness increase index F is then calculated according to Eq. 3 as a weighted sum of the scores (σ_i), with w_i being the correspondent weights:

$$F = \sum_{i=1}^{m} w_i \times \sigma_i; \sum_{i=1}^{m} w_i = 1 \tag{3}$$

Tab. 5.2: Definition of nontechnical aspects that increase attractiveness

ID	Definition	Reference
S1	Company ownership	[3]
S2	Presence of third-party highly attractive targets	[24]
S3	Presence of chemicals that can be used as WMD	[3]
S4	Past threat history	[3]
S5	Terrorist/activist activity in the area	[3]
S6	Political instability	[22]
S7	Ease in weapons gathering	[22]
S8	Local aversion due to company image and reputation	[23]
S9	Aversion due to local stakeholders engagement and awareness of technology	[23]
S10	Aversion for economic/environmental reasons and/or interactions with cultural/religious heritage	[23]

Tab. 5.3: Scores and weight associated with nontechnical aspects that increase attractiveness (see Tab. 5.2 for aspect ID definition)

Aspect ID	Weight (wi)	State	Description	Score (σ_i)
S1	0.0324	Presence	Public ownership/state participation in company management. Company may be seen as a symbol of state authority	1
		Absence	Private ownership	0
S2	0.1445	Presence	Presence of military targets, Institution buildings, embassies, monuments of high symbolic value, critical infrastructure in the site proximity.	1
		Absence	Absence of military targets, Institution buildings, embassies, monuments of high symbolic value, critical infrastructure in the site proximity.	0
S3	0.1445	Presence	Chemicals that can be used as weapons of mass destruction are stored/handled/processed/produced in significant quantities in the site.	1
		Absence	Chemicals that can be used as weapons of mass destruction are *not* stored/handled/processed/produced in significant quantities in the site.	0

Tab. 5.3: (continued) Scores and weight associated with nontechnical aspects that increase attractiveness (see Tab. 5.2 for aspect ID definition)

Aspect ID	Weight (wi)	State	Description	Score (σ_i)
S4	0.1692	Presence	Similar facilities or facilities owned by the same company have been object of previous attacks	1
		Absence	Similar facilities or facilities owned by the same company have never been object of attacks	0
S5	0.1445	Presence	Terrorist/activist groups are active in the area	1
		Absence	No terrorist/activist groups are active in the area	0
S6	0.0819	Low	A context of political stability and democracy exists. Governing authorities are legitimated and supported by populace.	0
		Medium	Few opposition groups willing to mine government authority exist and may be blamed for violent actions. Existence of political factions.	0.5
		High	Political instability and internal conflicts exist. Social order control and maintenance is periodically disrupted.	1
S7	0.0653	Low	Strict legislation concerning the transport, selling and detention of weapons of any nature. Effective and diffuse implementation of controls by police forces.	0
		Medium	Legislation concerning the transport, selling and detention of weapons is present but control is not a priority.	0.5
		High	Transport, selling and detention of weapons is poorly ruled and uncontrolled. Third-party interests favor the weapons market.	1
S8	0.0726	Low	Company reputation and image are extremely positive. Local community judge company activities beneficial.	0
		Medium	Company activities are accepted by local community. Few aversion motives of minor importance.	0.5
		High	Company reputation and image are extremely negative. Existence of organized aversion groups.	1
S9	0.0726	Low	High level of engagement of local stakeholders. Transparency and continuous information sharing to enhance community awareness of company activities.	0
		Medium	Medium level of engagement of local stakeholders. Company activities are accepted by local community. Few aversion motives of minor importance.	0.5
		High	No engagement of local stakeholders. Creation of a climate of suspicion and mistrust.	1
S10	0.0726	Low	No interactions with cultural/historical, archeological, religious heritage. Absence of activists groups on the area/no evidence of aversion by activist groups.	0
		Medium	No significant negative interactions with cultural/historical, archeological, religious heritage. Sporadic demonstrations of aversion by local activist groups.	0.5
		High	Negative interactions with cultural/historical, archaeological, religious heritage. Frequent demonstrations of aversion by activist groups attracting regional/national media attention.	1

The introduction of weights in Eq. 3 allowed accounting for the different degree of influence that incentives may have on adversaries' targeting logic. In order to limit the subjectivity in the computation of the criteria weights, the technique of analytic hierarchy process (AHP) was applied through the pairwise comparisons method [25].

In the specific case, the pairwise comparison matrix was built with a Saaty scale, which allowed converting from qualitative evaluations of the relative importance between two criteria to numbers ranging from 1/9 to 9. By applying AHP, the relative weights were calculated as the normalized values of the principal eigenvector of the pairwise comparison matrix; principal eigenvector elements are summarized in Tab. 5.3 in correspondence with each aspect considered.

As it can be seen from Tab. 5.3, the highest relative importance in increasing the perceived value of a facility, and thus attractiveness, was assigned to the existence of a past history of malevolent acts against the facility under analysis. Then the facility proximity to strategic targets, the storage and handling of weapons of mass destruction precursors and the confirmed presence of terrorist/activist cells in the area were considered as secondary aspects. Given the proposed scheme of scores and weights, the location specific "induction index" (ζ) spans over the range 1–2.

In order to check the consistency of the evaluations carried out through pairwise comparison, the *consistency index* (CI) and the *random index* (RI; e. g., the consistency index when the entries of the pairwise comparison matrix are completely random) were computed according to the rules described by Saaty [25], with the following results:

- $CI = 0.0136$
- $RI = 1.51$

The evaluated CI is more than one order of magnitude lower than the RI; this condition indicates that inconsistencies are tolerable, and a reliable result is expected from the AHP [25]. Therefore, the weighting system proposed for the scores of nontechnical triggers may be considered sufficiently robust for the present application.

5.3 Case studies definition

In the following, a set of case studies is defined in order to exemplify the proposed methodology. The layout adopted for the analysis is reported in Fig. 5.3. Three different plants are considered, in which flammable or toxic substances are stored and processed. The plants are sited near densely populated areas and vulnerability centers (see Section 5.2.2 for definition).

Two different hypothetical locations for the same plants were selected to show the relevance of a peculiar socio-political context in determining the perceived value of a potential terrorist target: Location 1 – Italy; location 2 – Libya. An overview of the political context, of the threat history and the existing aversion motives by activist groups or local stakeholders for the two locations in which the plants are sited was

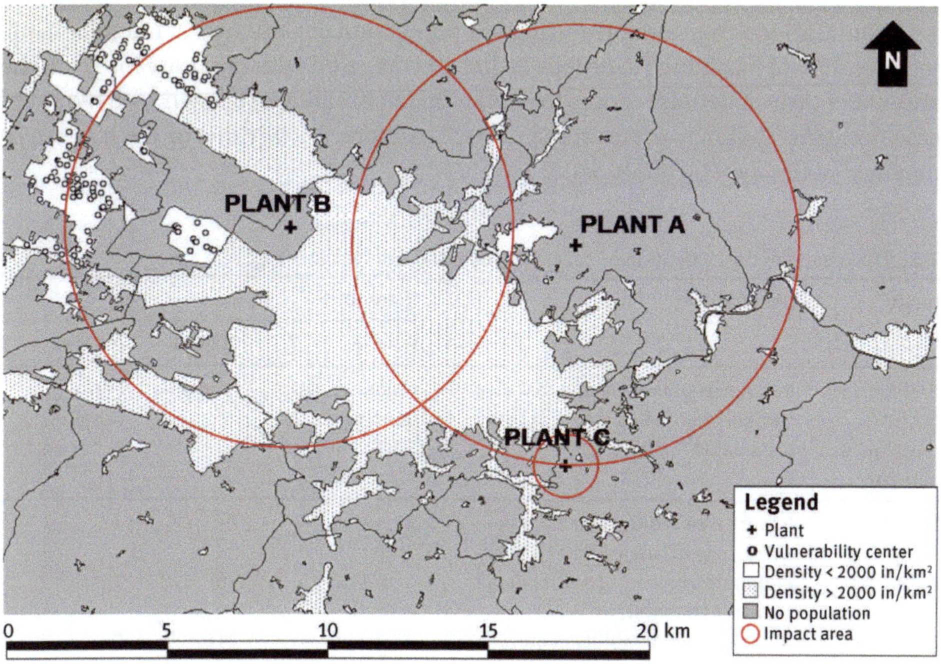

Fig. 5.3: Layout considered for the analysis of the case studies and impact area of the worst case accidents associated with the three chemical facilities considered (Plants A, B and C)

obtained. A survey was carried out, based on newspaper articles, public reports and declarations provided by operating companies and site managers. This allowed the attribution of the scores needed to determine the F index.

5.4 Results

Table 5.4 shows the results of the hazard based attractiveness index evaluation, following the procedure summarized in Fig. 5.2.

Plant A features relevant inventories of hazardous materials, leading to a hazardous substances index, I_{sub}, which is triple that of the other considered plants (e. g., B and C). This is aimed at penalizing the relevant severity of accidental scenarios associated with the plant. However, Plant A is not located in the proximity of urban areas and vulnerability centers, and then the associated territorial vulnerability index, I_{TV}, features intermediate values (see the scale shown in Fig. 5.2).

On the contrary, despite the fact that Plant B features a limited inventory of toxic substances, the proximity of densely populated areas and several vulnerability centers leads to high values of I_{TV}, resulting in a hazard based attractiveness of the site that is comparable to that evaluated for Plant A. In fact, for both plants a high value of hazard based attractiveness (I_H) is estimated, according to the metric summarized in Tab. 5.1.

Plant C is located in areas featuring low population density and, at the same time, the inventory of hazardous materials is limited compared with Plant A. Moreover, the absence of toxic materials reduces the potential impact area associated with the worst case accidents that may occur in the plant. Therefore, a low value of the index I_H is obtained in this case (see Tab. 5.4).

Tab. 5.4: Results of the evaluation of hazard based attractiveness index I_H

Input	**Description**	**Plant ID**		
		Plant A	**Plant B**	**Plant C**
Substance	Reference substance or substance category in the facility	Chlorine, petroleum products	Ammonia	Petroleum products
Substance category	Flammable (F)/toxic (T)	F & T	T	F
Impact	Impact area radius (km) based on worst case accident	7	7	1
Population	Population in the potential impact area (inhabitants)	225,832	386,538	1640
Vulnerability centers	Number of vulnerability centers	4	210	0
Index	**Description**	**Plant A**	**Plant B**	**Plant C**
I_{fl}	Flammable substance overall Index	640	0	50
I_{tox}	Toxic substances Index	50	30	0
I_{sub}	Hazardous substances Index	690	30	50
I_{FH}	Process facility hazard index	6	2	2
I_P	Population Index	3	3	2
I_{VC}	Vulnerability Centers Index	2	4	0
I_{TV}	Territorial Vulnerability Index	5	7	2
I_H	Hazard-based attractiveness Index	11	9	4
	Qualitative ranking (derived from Tab. 1)	HIGH	HIGH	LOW

In order to assess the influence of nontechnical triggers to the overall attractiveness of the three facilities considered, the effect of locating the plants in two different areas is accounted for through the procedure described in Section 5.2.3. The locations considered feature different political and social conditions. For Location 1 (i. e., Italy), a context of peace time is assumed, yet the absence of dedicated antiterrorism rules and practices is evidenced. Instead, in Location 2 (i. e., Libya), the context is one of political instability, and the documented presence of terrorist cells moved by political and religious motives is considered, as also confirmed by several news agencies (see specific references in [19]). For the sake of simplicity, private ownership of the companies for the three plants and the absence of strategic targets in the proximity of the plants are assumed.

The relevant aspects summarized in Tab. 5.2 are scored according to Tab. 5.3, and the results are shown in Tab. 5.5. The attractiveness increase index F is evaluated for the two locations supporting the calculation of the induction factor ζ. As shown in Tab. 5.5, the relevant terrorist threat featured by Location 2 causes a relevant increment of F. In fact, taking into account the documented presence of armed factions and the possible presence of terrorist cells, blamable for violent actions and generally for the unstable political situation in Location 2, the F index obtained is higher than the one calculated for Location 1 by almost one order of magnitude (see Tab. 5.5).

The combined evaluation of technical and nontechnical triggers allows the assessment of the overall attractiveness index I_A for each plant in the two locations. The results are summarized in Tab. 5.6, while Fig. 5.4 offers a graphical representation of the calculated I_A values, also showing the qualitative attractiveness ranking based on Tab. 5.1.

Tab. 5.5: Results of the evaluation of nontechnical triggers: overall attractiveness increase index (F) and induction index (ζ)

Aspect ID	Description	Weighed score	Location 1	Location 2
S1	Public Company ownership	$\sigma_1 w_1$	0	0
S2	Presence of third-party highly attractive targets	$\sigma_2 w_2$	0	0
S3	Presence of chemicals that can be used as WMD	$\sigma_3 w_3$	0	0
S4	Past threat history	$\sigma_4 w_4$	0	0.170
S5	Terrorist/activist activity in the area	$\sigma_5 w_5$	0	0.140
S6	Political instability	$\sigma_6 w_6$	0	0.082
S7	Ease in weapons gathering	$\sigma_7 w_7$	0.032	0.065
S8	Local aversion due to Company reputation	$\sigma_8 w_8$	0	0.073
S9	Aversion due to lack of local stakeholders engagement and awareness of technology	$\sigma_9 w_9$	0.036	0.073
S10	Aversion due to economic/ environmental reason and/or interactions with cultural heritage	$\sigma_{10} w_{10}$	0	0.036
F	Overall attractiveness increase index	-	0.068	0.639
ζ	Induction index	-	1.068	1.639

The relevant severity associated with the worst case accidents and the territorial vulnerability, for Plant A and Plant B, respectively, causes relevant values of overall attractiveness either in Location 1 or 2, thus without a relevant influence of nontechnical triggers on the qualitative ranking. In fact, a high value is obtained for Plants A and B (see Fig. 5.5 and Tab. 5.6) in both cases. This is due to the fact that the methodology penalizes the facilities which may lead to extremely severe accident scenarios follow-

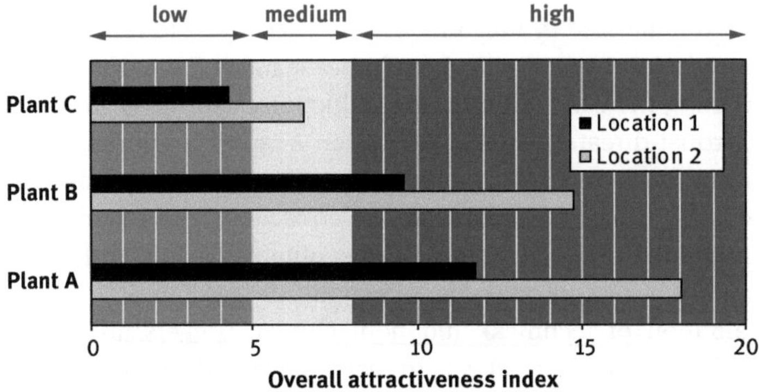

Fig. 5.4: Overall attractiveness assessment. Qualitative ranking is also reported, according to the classification summarized in Tab. 5.1.

ing an attack. In fact, it is considered that terrorists aim to cause as much damage as possible, and therefore, certain scenarios that would be considered extremely unlikely in the case of safety thinking, might actually be likely in the case of security thinking [26, 27].

Tab. 5.6: Overall results of attractiveness assessment for the case-studies

Parameter	Location 1			Location 2		
	Installation A	Installation B	Installation C	Installation A	Installation B	Installation C
I_H	11	9	4	11	9	4
ζ	1.07	1.07	1.07	1.64	1.64	1.64
I_A *(see Eq. 2)*	11.75	9.61	4.27	18.03	14.75	6.56
Overall attractiveness level	High	High	Low	Low	High	Medium

A different situation is obtained in the case of Plant C, for which "low" hazard based ranking is obtained, which leads to a "low" ranking in the case of Location 1, thus in absence of specific terrorist threats and activities. However, when performing a complete attractiveness assessment for Location 2, a more critical situation is obtained; it is ranked as "medium" due to the political context of the country housing the facility (see Fig. 5.5 and Tab. 5.6). Thus, despite the lower potentiality associated with the potential accidents in the facility, the evidence of local terrorist activities may as well contribute to the increment of the attractiveness of the facility.

5.5 Discussion

The analysis of the case studies exemplified the potentialities of the methodology in supporting the attractiveness assessment of different industrial facilities in several socio-political contexts. The methodology is a preliminary screening tool for security managers to prioritize the resources in the presence of "high" attractiveness results, to ask for more specific studies or to dispose of supplementary safety and security barriers able to protect a given process facility.

The results allowed highlighting, on the one side, the criticality associated with the severity of accidents triggered by terrorist attack in chemical facilities located in the proximity of vulnerable areas, and, thus, the importance of specifically addressing emergency planning, protection and prevention actions. On the other side, the importance of technical and socio-political aspects, as well as ideological and strategic incentives to an attack in the perspective of a holistic determination of target attractiveness were evidenced.

Concerning the limitations of this work, it is worth mentioning that the introduction of a metric based on scores which uncertainty boundaries may not be related to rigorous mathematical/physical models, but only to the different values and/or importance attributed to the different issues considered. Therefore, it clearly appears that expert elicitation of scores is of utmost importance for the quality of the results. Moreover, the outcomes of this approach need to be considered with the purpose of supporting a relative ranking of attractiveness associated with the different installations. In fact, independently from the numerical value obtained by the method, recognizing and identifying the security issues associated to a process facility is critical for decision-making processes [3, 9, 10, 16].

A limitation that may be object of future investigation is the assumption that there is no cross influence among the nontechnical aspects that contribute to attractiveness increase (see Tab. 5.2). This dependency was excluded for the sake of method simplicity, thus providing a schematization to drive the assignment of penalties related to socio-economic factors. On the one side, more complex elicitation techniques such as the weighted averaging models [28] or other aggregation techniques [29] may be adopted to define scores and associated weights. On the other side, the introduction of more sophisticated techniques, such as Bayesian networks and influence diagrams [30] may improve the analysis of interaction and cross influence among the scores.

Finally, it is crucial to highlight that the methodology developed may contribute the analysis of only a limited part of a complex SRA [16]. Nevertheless, determining the attractiveness of a process plant is aimed at supporting a preliminary screening to prioritize resources [9] and to determine further assessment needs in relation to the credibility of attacks on a given process facility in a particular context.

5.6 Conclusions

In the present study, a semiquantitative methodology for the assessment of industrial facilities' attractiveness with respect to malicious acts of interference was presented. The methodology considers two main aspects as targeting incentives. The first is related to the plant's hazard potential, i. e., the potential of causing severe damage to the population in the case of a successful attack leading to a major accident in the facility. The second aspect is the perceived value that a target may have for a specific threat, which is estimated considering the social, economic and political aspects associated with a plant and its location.

A set of case studies, drawn on the features of existing installations, was used to exemplify the procedure for attractiveness assessment. The attractiveness of different facilities, similar in type and hazard potential, was compared on the basis of the different socio-political and economical contexts.

The results allowed a preliminary ranking of the attractiveness of the targets. The methodology thus provides criteria to orient a more detailed analysis and provides criteria to prioritize emergency planning, protection and prevention actions.

References

[1] Baybutt P, Ready V. Strategies for protecting process plants against terrorism, sabotage and other criminal acts. Homel Def J 2003;2.

[2] Landucci G, Reniers G, Cozzani V, Salzano E. Vulnerability of industrial facilities to attacks with improvised explosive devices aimed at triggering domino scenarios. Reliab Eng Syst Saf 2015;143:53–62.

[3] Bajpai S, Gupta JP. Site security for chemical process industries. J Loss Prev Process Ind 2005;18:301–9.

[4] Ministère de l'Ecologie du Développement durable et de l'Energie. ARIA (analysis, research and information on accidents) database 2016. http://www.aria.developpement-durable.gouv.fr/?lang=en (accessed February 11, 2016).

[5] U. S. Department of Homeland Security. Chemical Facility Anti-Terrorism Standards (CFATS). Washington DC: U. S. Department of Homeland Security; 2007.

[6] European Commission. Council Directive, 2008/114/EC on the identification and designation of European Critical Infrastructures and the assessment of the need to improve their protection. Off J Eur Communities 2008;L345:75–82.

[7] European Commission. European Parliament and Council Directive 2012/18/EU of 4 July 2012 on control of major-accident hazards involving dangerous substances, amending and subsequently repealing council directive 96/82/EC. Off J Eur Communities 2012;L197:1–37.

[8] American Petroleum Institute, National Petrochemical & Refinery Association. Security Vulnerability Assessment Methodology for the Petroleum and Petrochemical Industries 2003.

[9] American Institute of Chemical Engineers – Center for Chemical Process Safety (AIChE-CCPS). Guidelines for Analyzing and Managing the Security Vulnerabilities of Fixed Chemical Sites. New York: American Institute of Chemical Engineers – Center of Chemical Process Safety; 2003.

[10] Störfall Kommission (SFK). SFK-GS-38 Report 2002.

[11] U. S. Department of Justice. A Method to Assess the Vulnerability of U. S. Chemical Facilities. Report NCJ 195171. Washington DC: Office of Justice Programs; 2002.

[12] Jochum C. Can chemical plants be protected against terrorist attacks? Process Saf Environ Prot 2005;83:459–62.

[13] Uth H-J. Combating interference by unauthorised persons. J Loss Prev Process Ind 2005;18:293–300.

[14] Bajpai S, Gupta JP. Terror-proofing chemical process industries. Process Saf Environ Prot 2007;85:559–65.

[15] Post JM, Ruby KG, Shaw ED. The Radical Group in Context: 1. An Integrated Framework for the Analysis of Group Risk for Terrorism. Stud Confl Terror 2002;25:73–100.

[16] American Petroleum Institute (API). ANSI/API Standard 780 – Security risk assessment methodology for the petroleum and petrochemical industry. Washington DC: American Petroleum Institute; 2013.

[17] Jaeger CD. Chemical facility vulnerability assessment project. J Hazard Mater 2003;104:207–13.

[18] FEMA Federal Emergency Management Agency. FEMA 452 – A How-To Guide to Mitigate Potential Terrorist Attacks Against Buildings. New York, NY: Federal Emergency Management Agency; 2005.

[19] Argenti F, Landucci G, Spadoni G, Cozzani V. The assessment of the attractiveness of process facilities to terrorist attacks. Saf Sci 2015;77:169–81.

[20] Argenti F, Landucci G. Advanced attractiveness assessment of process facilities with respect to malevolent external attacks. Chem Eng Trans 2016;53:133–8.

[21] Landucci G, Tugnoli A, Spadoni G, Cozzani V. LNG regasification terminals: Assessment of accidents due to external acts of interference. 11th Int. Probabilistic Saf. Assess. Manag. Conf. Annu. Eur. Saf. Reliab. Conf. 2012, PSAM11 ESREL 2012, vol. 6, 2012, p. 4373–82.

[22] Kis-Katos K, Liebert H, Schulze GG. On the origin of domestic and international terrorism. Eur J Polit Econ 2011;27.

[23] Pape RA. The Strategic Logic of Suicide Terrorism. Am Polit Sci Rev 2003;97:343–61.

[24] Ackermann G, Abhayaratne P, Bale J, Bhattacharjee A, Blair C, Hansell L, et al. Assessing Terrorist Motivations for Attacking Critical Infrastructure. Monterey, CA, USA: Center for Non-Proliferation Studies, Monterey Institute of International Relations; 2007.

[25] Saaty TL. How to make a decision: The Analytic Hierarchy Process. Eur J Oper Res 1990;48:9–26.

[26] Reniers GLL, Dullaert W, Audenaert A, Ale BJM, Soudan K. Managing domino effect-related security of industrial areas. J Loss Prev Process Ind 2008;21:336–43.

[27] Reniers GLL, Audenaert A. Preparing for major terrorist attacks against chemical clusters: Intelligently planning protection measures w. r. t. domino effects. Process Saf Environ Prot 2014;92:583–9.

[28] Cooke RM, Goossens LJH. Procedures guide for structured expert judgment. Report EUR 18820 En. Brussels, Belgium: Commission of European Communities Directorate; 1999.

[29] Boring RL. A review of expertise and judgment processes for risk estimation. Proc. Eur. Saf. Reliab. Conf. 2007, ESREL 2007 – Risk Reliab Soc Saf 2007;2:1901–7.

[30] Khakzad N. Application of dynamic Bayesian network to risk analysis of domino effects in chemical infrastructures. Reliab Eng Syst Saf 2015;138:263–72.

Laobing Zhang, Genserik Reniers

6 Applying game theory for adversarial risk analysis in chemical plants

6.1 Introduction

The 9/11 attack in New York reminded researchers in academia as well as managers in industry to pay more attention to the protection of critical infrastructures from intentional attacks. Chemical plants, due to their importance for citizen's daily lives and societal welfare and due to their vulnerabilities to terrorism, are listed in the 13 critical infrastructures by the American Department of Homeland Security [1]. However, a lack of research on how to protect chemical plants from terrorists still exists. Reniers et al. [2], among others, define the risk caused by intentional events as "security risk," while the risk caused by nonintentional events can be categorized as "safety risk."

Two well-known methodologies have been developed for assessing security risks. In 2002, the so-called security risk factor table (SRFT)[3, 4] was first proposed by the Advanced Chemical Safety Company to carry out a security risk assessment for a given facility. The basic idea of this method is to identify the security-related factors of the facility, rate them on a scale from 0 to 5, with 0 being the "lowest risk" and 5 being the "highest risk", then sum up the score of each factor to measure the security risk status of the facility. In 2003, the American Petroleum Institute (API) published a Security Risk Assessment framework (API SRA framework) [5] for petroleum and petrochemical industries. In the API SRA framework, security risk is defined as a function of consequences and likelihood; likelihood is a function of attractiveness, threat, and vulnerability. The API SRA framework is a systematic process evaluating the likelihood that a threat against a facility would be successful, and considering the potential severity of consequences to the facility itself, to the surrounding community, and to the supply chain. This method is performed qualitatively using the best judgment of the SRA Team. Compared to SRFT, the API SRA framework is more concrete to execute, and not only considers the facility itself, but also its surroundings.

Since being published, the above mentioned methodologies have been extensively conducted in industrial practice. However, there are criticisms of the methodologies themselves. Without modeling the intelligent interactions between the defenders and security adversaries, these methodologies might lead to misallocation of limited security resources [6, 7]. For instance, the API SRA framework models the threat's attractiveness of attacking the targets in a probabilistic way, without considering that the implementation of countermeasures on targets would change the targets' vulnerability as well as their attractiveness. Nonetheless, game theory would

https://doi.org/10.1515/9783110499087-006

be good at modeling the strategic interactions among intelligent players. As stated by Cox [8], combining game theory and conventional (probabilistic) risk analysis techniques is very promising for security risk research. Conventional risk analysis techniques (e. g., the API SRA framework) can provide quantitative inputs for game theory models, while game theory models could process these inputs in an intelligent way, making the best use of available data.

Research on computational game theory [9] enables researchers to solve large-scale games, resulting in game theory applications in various domains. Game theoretic risk analysis has been sufficiently studied in a number of domains (e. g., network security [10], airport protection [11], etc.). However, in the process industry, only a few game theoretic studies have been published. Zhang and Reniers [12] proposed a non-zero-sum, complete information, simultaneous game to improve chemical security by optimally setting the security alert levels (SAL) at different "typicals". Pavlova and Reniers [13] studied how to form maximal security cooperation with minimal expense in chemical clusters, by employing a two-stage cooperative game. Talarico et al. [14] developed the so-called MISTRAL game for improving the security of a multi-modal chemical transportation network, thereby considering both simultaneous and sequential solutions.

In the remainder of this chapter, in Section 6.2, the general intrusion detection approach in a chemical plant is introduced. In Section 6.3, the security game in general is briefly illustrated, while an application of the security game within the chemical industry (i. e., the CPP game) is given in Section 6.4. Section 6.5 discusses three basic equilibrium concepts of the CPP game. A discussion of a further extension of the CPP game is given in Section 6.6, and conclusions are drawn in Section 6.7.

6.2 General intrusion detection approach in a chemical plant

Figure 6.1 shows a general physical intrusion detection approach in a chemical plant. The different layers of "perimeter" divide the plant area into different layers or "zones". In order to intrude higher level zones, a potential intruder must pass the lower level zones first. A further realistic assumption is that an intruder would never come into the same level of zones twice. For example, if an environmental activist aims to shut down a facility in ZONE 2_2, he would have to pass ZONE 0, perimeter 1, ZONE 1_1, and perimeter 2, while the assumption is present that he would not step into ZONE 2_1 because otherwise he would come twice into zone levels 1 and 2. The security countermeasures at the perimeters (e. g., access control, checkpoint, etc.) and the zones (e. g., patrolling, safety barriers, etc.) make the targets in higher zones more secure (i. e., less vulnerable). A "typical" in the system is defined as the summation of items constituting a security barrier and thus describes the specific detailed characteristics of a security measure installed at a plant or at a part thereof [15]. For example, at the Main Entrance in Figure 6.1, there could be some security staff, barriers, identity recognition system, etc., all of which constitute a security barrier, thus the Main Entrance is a "typical".

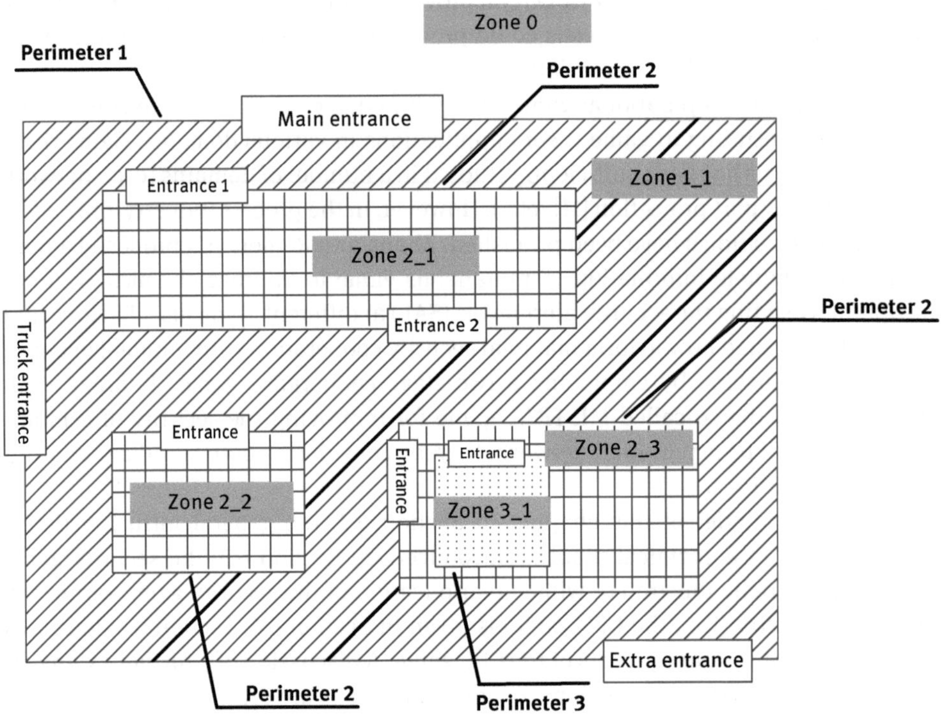

Fig. 6.1: General intrusion preventing in process plants (source: Zhang and Reniers [12]).

Figure 6.2 illustrates the potential attackers' behavior. The attackers would, firstly, decide which target to attack, secondly, choose the "easiest way" (the so-called critical path) to reach the target, and thirdly, decide on the attack scenario (e. g., armed or unarmed). Note that these are not independent steps, e. g., when the attackers choose the target, the difficulties of reaching the target is of course a very important factor for them to consider. Formula (1) gives the probability of successfully reaching the target.

$$P = \prod_{i=0}^{I} P_i^z \cdot \prod_{j=1}^{I} P_j^p . \tag{1}$$

In which the I denotes the zone level number of the target, for example, in Figure 6.1, if a target located in ZONE 3_1, then we have $I = 3$. P_i^z denotes the probability of successfully passing the zone level i; P_j^p denotes the probability of successfully passing the perimeter j.

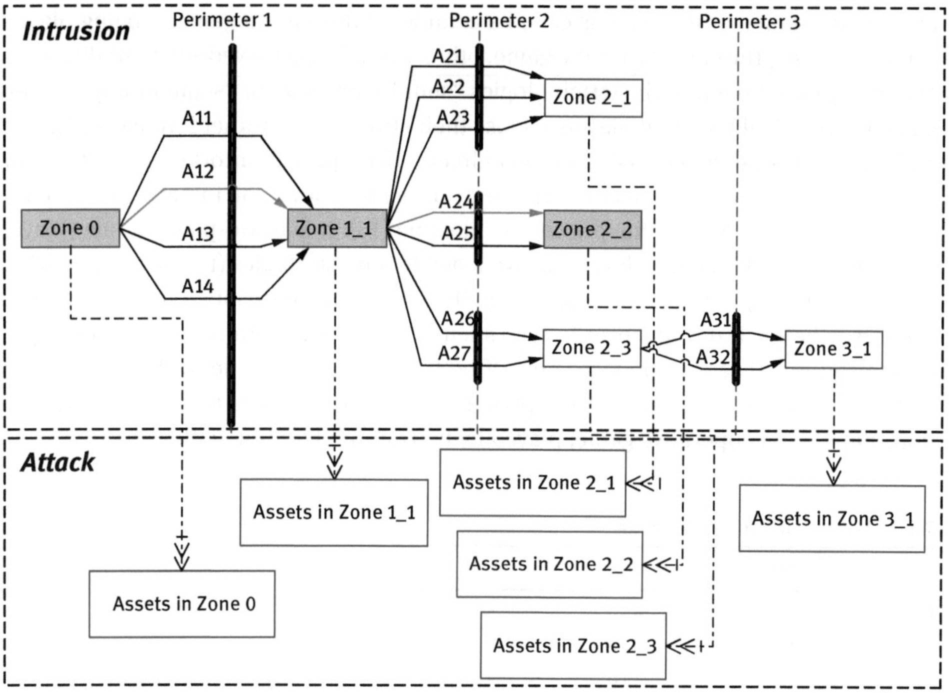

Fig. 6.2: The intrusion and attack procedure (Source: Zhang and Reniers [12]).

6.3 Security game

In this section, some basic knowledge of game theory is given, followed by the definition and explanation of the so-called security game.

6.3.1 Game theory

Game theory was created to model intelligent interactions among strategic agents. Classical game theory research is usually based on two assumptions: A1) rational players, which means that each player aims at optimizing his own payoffs and players know that others also aim at optimizing their own payoffs, and so forth; and A2) common knowledge of the game, which means the rules of the game, the players' feasible strategies, the players' payoffs under certain strategy tuples are common knowledge among all the players. A game theoretic model consists of three components: C1) players; C2) strategy sets; and C3) payoff function. In a game, if it is possible to have a contract among the players, the game is called a cooperative game, otherwise, it is called a non-cooperative game; if the assumption A2 does not hold, the game is called an incomplete information game, otherwise it is called a complete information game; if being played simultaneously, the game is a static game, otherwise if being played sequentially, it is a dynamic game. Note that the terminologies "simultaneously" or "sequentially" do not mean the physical time that players choose their strategy, but rather it reflects whether a player knows other players' decisions already (i. e., perfect information) when he/she acts, or not (i. e., imperfect information). Table 6.1 shows the famous case in classical game theory, that is, the prisoner's dilemma [16]: it is a two-player game (Player 1 = P1, Player 2 = P2); each player has two feasible pure strategies (to betray each other or to keep silent); if both choose to betray, they would be imprisoned for 3 years, if P1 chooses to betray but P2 chooses to keep silent, then P1 will be freed and P2 will be put into prison for 10 years, and vice versa. If both choose to keep silent, they will be put into prison for 1 year. Generally, the prisoner's dilemma game is a complete information, static, noncooperative game, played by two definitely rational players.

Tab. 6.1: I. The prisoner's dilemma

	P2		
P1		B	S
	B	−3,−3	0,−10
	S	−10,0	−1,−1

Although there are many economists and mathematicians who are doing, or have been doing, research on game theory, three milestones should be emphasized. The first is John von Neumann's work [17], proposing the formalism of game theory. Before John von Neumann, there was some scattered research that already used the ideas of game theory, such as Augustin Cournot's work, James Madison's work, etc. However, John von Neumann was the first to formalize the different studies on game theory, and after his work game theory became a unique field. The second is John Nash's work in 1950 [18]. Nash proved that in a finite game (finite players and finite strategies), an equilibrium always exists, and he proposed the very famous concept in game theory, namely, the "Nash equilibrium". The third is John Harsanyi's work [19], that systematically studied the incomplete information game, and proposed the famous concept "Harsanyi transformation", which is able to transfer an incomplete information game to a complete but imperfect information one.

6.3.2 Security game

Since first formalized by John von Neumann in 1940, game theory has been extensively used in various domains, such as micro economics, political science, computer science, biology, etc. Over the past few decades, security scientists (cyber security/

information security/network security/physical security or critical infrastructure protection) have started to employ game theory as a methodology to assess security risks from intelligent and strategic attackers. A special type of game, named the Bayesian Stackelberg game (BSG) [20], has been emphasized frequently in physical security, due to its capability of modeling uncertainties (Bayesian) and sequential moves [21].

Among the game theoretic security models, although some authors model their game as a simultaneous one [12, 22], more authors would agree that in security, games should be sequential due to the fact that defenders must always move first (leader) and the attacker follows (follower), taking into consideration the strategic observation of the defender's move. Heinrich von Stackelberg [21] first studied the leader–follower game in 1934, so these types of games are called "Stackelberg games". In fact, the famous attacker–defender model [1], where an attacker acts rationally based on the observation of a defender's action and aims at maximizing his own payoff, and the defender knows this and acts accordingly, is analogous to a Stackelberg game.

It is acceptable to say that an attacker could have full information about the defender's side, since some studies have pointed out that "using [public sources] openly and without resorting to illegal means, it is possible to gather at least 80% of information about the enemy" [1, 23], and "in fact, we find that public sources often provide 100% of the information required to plan a devastating attack on an infrastructure system" [1]. Unfortunately, it is difficult to say that the defender also has complete information about the attackers for the following reasons: R1) The defender faces different types of attackers, such as disgruntled employees, terrorists, and environmental activists. Different types of attackers have different goals, behaviors, and rationalities. R2) Even one type of attacker's[1] utility function is difficult to estimate. For example, in some research, the authors assume that the defender's loss is equal to the attacker's gain, however, due to different psychological behaviors, defenders and attackers might value the same asset in a different way. R3) Although attackers could acquire almost 100% of the information required for a plan, the defender could not be sure if the attacker really has 100% of the information. Some research also points out that under certain conditions, the defender could have "first-mover advantage" [24], thus she can make the information public to make sure the attackers have 100% of the information. Unfortunately, these conditions are not always satisfied. R4) There are not enough historical data on security risk, making the estimation of the attackers even more difficult. In order to meet these uncertainties of attackers, a Bayesian framework is needed.

The previous two paragraphs illustrate the reason why the Bayesian Stackelberg game is popular in security research. Figure 6.3 shows an example of a Bayesian Stackelberg game in extensive form. In this game, we assume that the defender has two assets (A1/A2), but she can only protect one (i. e., limited budget/resources

1 In security game terminology, the defender is normally represented as she/her/her, while the attacker is denoted as he/him/his.

assumption) at the same time at cost 2. The attacker would choose one asset to act based on the perfect observation, here we assume that there are two types of attacks: Type I is theft and type II is sabotage, with the same probabilities. To simplify the question, we assume that if the attacker acts on the protected asset, he would fail and if he acts on the unprotected asset, he would succeed. For the defender, if the attacker acts successfully on A1/A2, she would lose 3 and 4, respectively. For the theft, what the defender loses is assumed to be what the attacker gains, so his gains of a successful action on A1/A2 are 3 and 4, respectively. For the sabotage, what the defender loses is not what he gains, but relations exist between both. We assume what the attacker gains is half of the defender's loss, i. e., 1.5 and 2 for A1/A2, respectively. In Figure 6.3, if the defender chooses to protect A1, and the type I chooses to attack A1, in this case, the attacker would fail, thus the defender only loses the protection cost (i. e., -2), and the attacker gains nothing (i. e., 0, it is worth noting that in this simplified illustrative example, the attack costs are not considered). Conversely, if the defender chooses to protect A1, but the attacker type I chooses to attack A2, in this case, the attacker would succeed, thus the defender would lose the protection cost as well as the value of A2 (i. e., $-(2+4)=-6$), and the attacker would gain the value of A2 (i. e., 4). Other payoffs shown in Figure 6.3 can be calculated analogously. In this Bayesian Stackelberg game, the defender must face the uncertainties of the attacker types, thus when she solves the game, she must consider the two extensive forms together.

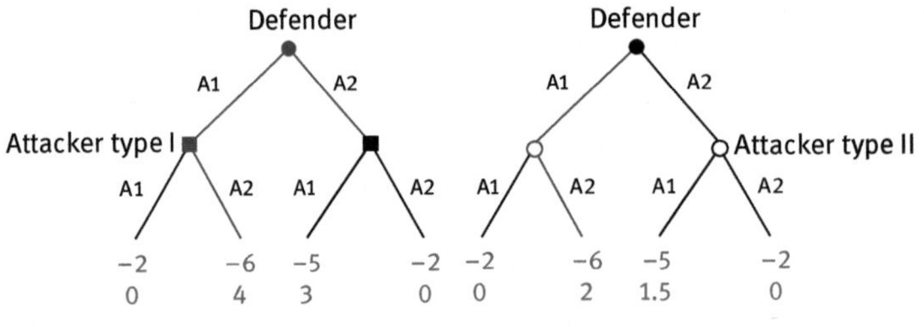

Fig. 6.3: Bayesian Stackelberg game.

Tambe evaluated a type of Bayesian Stackelberg game, in which what is better for one player is worse for the other player, as a "security game" in his book [20]. This characteristic might mislead readers to consider the security game as a zero-sum game. However, the security game is in reality not a zero-sum game for the following reasons: S1) Attacker and defender might value the same asset differently. For example, human lives may always be the most important asset for the defender, but this may not be the case for a terrorist attacker, or some important documents are really valuable to the defender but are useless for the thieves, etc. S2) Defense and adversarial actions cost money [25]. When we calculate the payoffs of the security game, defense and adversarial action costs should always be taken into consideration. These costs show likely no zero sum properties, and the defender cannot gain anything from the attack cost and vice versa.

6.4 Chemical plant protection game (CPP game)

If the general security game is applied to security risk analysis related to the physical intrusion detection approach in a chemical plant, we obtain the chemical plant protection game (CPP game). A game theoretic model mainly consists of three parts: the players, the strategy sets, and the payoff matrix. These three parts will be discussed in the following sections.

6.4.1 Players of the CPP game

In the CPP game, there are only two players: one is the defender and the other is the potential attacker (whereby different types are possible). Questions that are valuable for discussion here are whether the players are rational and whether they have the complete information about the game.

For the first question, whether the players are rational, many papers state that attackers will analyze defenses and then plan accordingly. They can thus be assumed to be rational. The defenders, who in our case are the security managers in chemical companies and have abundant experience and are highly educated, can also be assumed to be rational. Security experts might argue that some types of attackers, e. g., suicide bombers, are difficult to be assumed as rational players. Nevertheless, a rational player in a game theoretic model implies that he/she is motivated by maximizing his/her own payoff [26]. Therefore, for dangerous attackers (e. g., terrorists), if we model their payoff as the defender's losses, and we consider the intuition that attackers are aiming at maximizing the damage, then they can be assumed as rational player as well.

As regards the second question, whether players have the complete information about the game, it is difficult to assess. Since the attackers preferably will collect extensive security information about the plant, and some studies show that based on the media, terrorists can collect at least 80% (sometimes even 100%) of the information that is needed to execute a successful attack [1], it may indeed be reasonable to assume that the attackers have the complete information about the game, including the strategy sets of both themselves and the defender, as well as the payoffs when both players choose certain strategies. Unfortunately, it is difficult to say whether defenders have the complete information about the security game, since to date, we do not have enough knowledge about the attackers. A possible solution to face this shortage is the Bayesian game. In a Bayesian game, we can define different types of attackers as well as different attack scenarios. For each of them, we can estimate their occurrence probabilities.

For a more detailed discussion of "rationality", "complete information", and also common knowledge of a security game, the interested readers are referred to Guikema [27]. Section 6.6 in this chapter discusses some concepts and algorithms dealing with the cases when the rationality and common knowledge assumptions do not hold.

6.4.2 Strategy Sets of the CPP Game

There are three options of attack in a chemical plant: intrusion, inside job, and a remote attack. The focus of this chapter is intrusion, since this is the most visible part of anti-terrorism security policies in a chemical plant. Most chemical plants are also focused on this type of security in current industrial practice. Attacks from insiders, as well as remote attacks (e. g., with drone technology) are of a different nature, and require their own research activity.

To secure a chemical plant, security managers need to decide where to implement each typical at what security alert level. To execute an attack, the attacker needs to decide which target to attack first and then choose the critical intrusion path and attack scenario.

The defender's strategies are given by:

$$S_d = \{s_1, s_2, \ldots, s_n\} \tag{2}$$

With:

$$s_i = z^0 \times \prod_{r=1}^{Q} \left(A_1^r \times A_2^r \times \ldots \times A_{ent(r)}^r \times z_1^r \times z_2^r \times \ldots \times z_{sub(r)}^r\right) \tag{3}$$

In which: S_d: defender's strategy set; s_i: a specific defense decision, also called "pure strategy" in game theoretic terminology, of the defender; n: number of pure strategies of the defender; z_i^r: detection level in i^{th} subzone in zone level r; A_i^r: detect level at the ith access of perimeter r; $ent(r)$: number of accesses in perimeter r; $sub(r)$: number of subzones in zone level r; Q: total zone levels in the plant; and $\times$: the cross product.

Assume further that the chemical plant can set each typical to k_d different security alert levels, for example, $k_d = 3$ or $k_d = 5$, then we can compute the number of possible defender strategies using the following formula:

$$n = k_d^{1+\sum_{r=1}^{Q}(ent(r)+sub(r))} \tag{4}$$

Furthermore, the attacker's strategies are:

$$S_a = \{s_1, s_2, \ldots, s_m\} \tag{5}$$

With:

$$s_i = a \times \prod_{r=1}^{I} A_{j_r}^r \times e \tag{6}$$

In which:S_a: attacker's strategy set; s_i: a specific attack action, also called "pure strategy" in game theoretic terminology, of the attacker; m: number of pure strategies of the attacker; a: the target asset; $A_{j_r}^r$: the j_r access to pass the rth perimeter; e: attack scenario.

For simplicity, we can see the attacker's pure strategy as which asset to attack, how to intrude to reach the asset, and with what attack scenario.

The number of pure strategies of the attacker can be computed by employing the following formula:

$$m = k_a \cdot \sum_{r=1}^{Q} \sum_{i=1}^{sub(r)} \left(\tau_i^r \cdot \prod_{j=1}^{r} ent\,(j)\right) \tag{7}$$

Where k_a is the number of different attack scenarios, and τ_i^r denotes the number of targets located in the ith subzone in zone level r.

According to formulas (4) and (7), we realize that n and m will increase dramatically as the scale of the plant grows. However, in practice, limited budgets/resources are available for both defenders and attackers, thus n and m can be reduced, see, for example, Talarico et al. [14].

For the convenience of expressing the players' strategies, a code protocol that can map to the strategies as well as the strategy index is needed. Since the defender's strategies depend on the security alert level at each typical, we can easily code her strategies as a series of numbers $d_1 d_2 \ldots d_{T_n}$, where d_i equals the detection level at the ith typical and T_n indicates the total number of typicals (we also have $T_n = 1 + \sum_{r=1}^{Q}(ent\,(r) + sub\,(r))$). We also code the attacker's strategy as a series of numbers by using the code protocol shown in Figure 6.4.

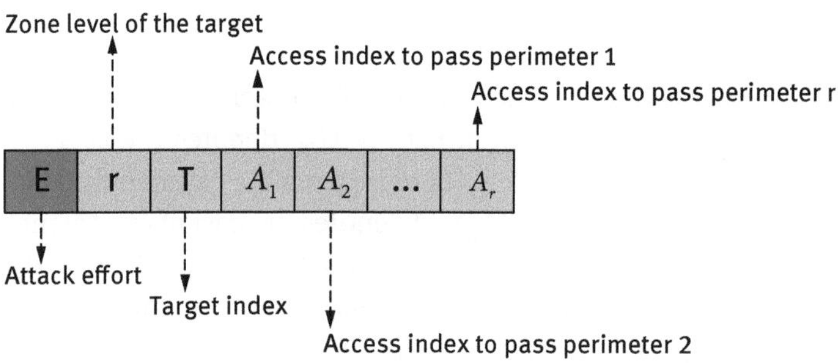

Fig. 6.4: Code protocol of attacker's strategies.

For instance, the defender's strategy code "1111 2222" means that there are eight typicals in the plant, and the detection levels at the first four typicals are "1" while at the later four typicals are "2". As an example of an attacker's strategy code, a suicide bomber aims to attack target τ located in zone 2_2 in Figure 6.1, and he chooses the intrusion path as the red (bold) line in Figure 6.2; this strategy can be represented as (suicide bomb,2,τ,2,4), as defined in Figure 6.4.

6.4.3 Payoffs of the CPP Game

In any game, payoffs are numbers that represent the motivations of players, for instance, profit, quantity, "utility," or other continuous measures (cardinal payoffs), or may simply rank the desirability of outcomes (ordinal payoffs). Usually, a payoff should be some function of the strategy tuple played by all players.

The intruder aims at causing some damage (or steal something, etc.) to the target, as analyzed in the previous section, and to do so, he first should choose a target to reach and then plan an intrusion path. After reaching the target, he will attack the target, with a certain probability of success due to the difficulties of having to surpass the defenses related to the target. Irrespective of the fact whether the intrusion and the attack are successful, an attack has a financial aspect attached to it (that is, it will cost money), with the exact amount depending on the scenario used.

The defender aims at preventing losses from an attack. If she does not invest in security, the intruders can intrude more easily into the plant and execute a successful attack. In this chapter, we assume that security investments are only used to reduce the probability of successful intrusion.

Based on the above analysis, the attacker's and defender's payoff can be calculated by the following two formulas:

$$u_a(s_a, s_d) = \tilde{P}(s_a, s_d) \cdot \tilde{P}_y(s_a) \cdot \tilde{L}(s_a) - C_a(s_a) \tag{8}$$

$$u_d(s_a, s_d) = -(P(s_a, s_d) \cdot P_y(s_a) \cdot L(s_a) + C_d(s_d)) \tag{9}$$

In which, P ($\tilde{P}$) denotes the successful probability that the attacker can reach the target (see the definition in formula 1, in Section 6.2); P_y ($\tilde{P}_y$) denotes the successful probability that the attack will be executed; L ($\tilde{L}$) represents the estimated consequence of a successful attack on the target; and C_d (C_a) denotes the cost of the defence (attack) plan; from the defender's (attack's) perspective. Note that in this chapter, the exit procedure of the attack is not considered, in order to simplify the model.

The definition of the strategy sets implies that the CPP game has finite strategies for each player, and formulas (4) and (7) give the number of strategies for the defender and attacker, respectively. Implementing formulas (8) and (9) for each tuples of defender–attacker strategies, we finally have the payoff matrix, denoted as U_a, U_d.

6.4.4 Discussion of the parameters

To calculate the payoff of the game by formulas (8) and (9), the following parameters are needed: the successful passing probabilities at each typical for each security alert level as well as for each attack scenario (i. e., p_i^z, p_j^p in formula 1); the conditional probability of an attack (i. e., P_y); the estimated consequences and gains (i. e., L); and the defence costs and attack costs (i. e., C).

The successful passing probability at a typical depends on the security alert level at the location and the attacker's intrusion scenario. For instance, at the main entrance of a chemical plant, if there would be a x-ray baggage scanner, and the attacker chooses the attack scenario as bringing a bomb into the plant, in this case the passing probability at the main entrance would be very low, while if the attacker chooses the attack scenario as intentional misoperation on some installations (which can also be very dangerous), then the passing probability at the main entrance would be higher than the former scenario.

The conditional probability of an attack and the estimated consequences and gains are target specific and scenario specific parameters. For instance, if the vehicle borne improvised explosive device (VBIED) scenario is executed on a tank farm of a refinery, the probability and the consequence would both be high, while if a thief would like to steal a technique document from the administrative building, the probability would be lower and the consequence would be either low or high.

The computation of defense and attack costs depends on the security alert level and the attack scenario. For example, in the scenario of a suicide bomb attack, the bomber's cost C_a contains the cost of the bomb, the bomber's life, the vehicle he uses to intrude the plant, and other costs. Furthermore, different types of bombs, different bombers (well-trained or not), different vehicles related to different attack scenarios, and a different scenario related to a different attack cost, can be conceptualized and mapped.

It is worth noting that all these parameters should be estimated by expert teams, e. g., the SRA team defined in the API SRA framework [5]. Furthermore, as also indicated in formulas (8) and (9), all these parameters should be estimated from both the defender's and the attacker's perspective. For example, for the probability of passing a typical, the expert team needs to estimate the defender's estimation of the probability (i. e., P) as well as the defender's estimation of the attacker's estimation (i. e., $\tilde{P}$). In conventional security risk analysis methods, e. g., the API SRA framework, these parameters are not separated. This implies that, in these methods, the defender thinks that the attacker has the same valuation as the defender, which is not always true. For instance, for the probabilities of passing a typical, a risk-seeking attacker might always have higher estimation than a risk-neutral defender. Furthermore, for a successful attack, the attacker's gain is not necessarily equal to the defender's loss. For a further discussion of this iterated estimation of parameters, interested readers are referred to Rios and Insua [28].

6.5 Solutions of the CPP game

A solution of a game is a pair (not necessarily unique) of strategies, which is the so-called pure strategy solution, or a probability distribution of each of the strategies, which is the so-called mixed strategy solution, that a rational pair of players might use. In this section, we first introduce the mixed strategy concept in game theory, then three primary solution concepts are introduced: the Nash equilibrium; the Strong Stackelberg equilibrium; and the Bayesian Stackelberg equilibrium.

6.5.1 Mixed strategy

A pure strategy is a specific move/action/decision that a player will make or carry out. The strategy sets defined in formulas (3) and (6) are pure strategies for the defender and the attacker, respectively.

A mixed strategy is a strategy consisting of possible moves and a probability distribution (collection of weights) which corresponds to how frequently each move is to be played [29]. For example, a mixed strategy for the defender can be represented as:

$$y = (y_1, y_2, \ldots, y_n)^T$$

Where $y_j \in [0, 1]$, and $\sum_{j \in N} y_j = 1$. y_j denotes the probability that the ith pure strategy is played by the defender. Similarly, an attacker's mixed strategy can be defined as:

$$x = (x_1, x_2, \ldots, x_m)^T$$

Where $x_i \in [0, 1]$, and $\sum_{i \in M} x_i = 1$.

In the above definitions, we have $N = \{1, 2, \ldots, n\}$, $M = \{1, 2, \ldots, m\}$. Furthermore, we represent the defender's mixed strategy space as Y, while the attacker's as X.

When a mixed strategy is used, the players' payoffs should be computed as the expected payoffs under certain probability distributions on pure strategies,

$$v_d(x, y) = x^T \cdot U_d \cdot y \tag{10}$$

$$v_a(x, y) = x^T \cdot U_a \cdot y \tag{11}$$

Mixed strategy is a theoretic terminology in the game theory domain. If applied to the CPP game, a mixed strategy for the defender can be explained as the defender playing different pure strategies from day to day, according to mixed probabilities. A mixed strategy for the attacker can be explained as the attacker playing pure strategies whose probabilities are not 0, but the defender could not know which exact pure strategy the attacker would use.

Another valuable question about mixed strategy is whether the defender could play different pure strategies from day to day. For fixed equipment (e. g., a face recognition machine), it is usually not convenient to move it among different typicals in order to set the different security alert levels at different days. Hereby, mixed strategies can only be used for mobile equipment (e. g., human guards etc.), or they can be used for determining how frequently to use the fixed/mobile equipment (notice that in practice, some equipment is not always in a working condition/state, since it might be periodically shut down in order to reduce operation cost, for instance).

6.5.2 Nash equilibrium

In the case that the attacker cannot observe the defender's implemented strategies, the CPP game would be a simultaneous move game. Note again that the defender always must implement her defense strategy first, thus the terminology "simultaneous" here does not mean they move simultaneously on the physical time axis, but it means that when the attacker moves, he does not know what the defender's implemented strategy is.

In a simultaneous game, neither player knows each other's implemented strategy, thus both of them play their own best responses to each other's strategy. In this case, the Nash equilibrium can be used to predict the outcome of the game.

A Nash equilibrium (for two players) is a solution where a change in strategies by either player will lead him to obtaining less payoff if the other player remains with his current strategy, or, in other words, a Nash equilibrium point is a stable point of the players' strategies, where no one has motivations to change strategy.

A pure strategy Nash equilibrium for the CPP game can be defined as (s_a^*, s_d^*) such that:

$$u_d(s_a^*, s_d^*) \geq u_d(s_a^*, s_d), \quad \forall s_d \in S_d \tag{12}$$

and

$$u_a(s_a^*, s_d^*) \geq u_a(s_a, s_d^*), \quad \forall s_a \in S_a \tag{13}$$

A mixed strategy Nash equilibrium of the CPP game is a pair of strategies (x^*, y^*) such that

$$v_d(x^*, y^*) \geq v_d(x^*, y), \quad \forall y \in Y \tag{14}$$

and

$$v_a(x^*, y^*) \geq v_a(x, y^*), \quad \forall x \in X \tag{15}$$

For algorithms to find Nash equilibria, interested readers are referred in case of pure strategy Nash equilibria to Gibbons [30] and for mixed strategy Nash equilibria to Lemke and Howson [31].

6.5.3 Strong Stackelberg equilibrium

A more reasonable idea is that the attacker would be able to know the defender's implemented strategies, when he moves. Security disaster reports (e. g., the 9/11 report) more and more frequently show that our adversaries are intelligent, and they behave accordingly. Therefore, it is necessary to think about what if the attacker has enough intelligence about the defense plan. In this case, the CPP game is played sequentially, or we call it a Stackelberg CPP game.

In the Stackelberg CPP game, the defender moves first (i. e., forming the game leader), followed by the attacker (i. e., forming the game follower), with full observation (i. e., the follower has perfect information about the game).

A strong Stackelberg Equilibrium (SSE) $(\overline{k}, \overline{y})$ for the Stackelberg CPP game is defined by formulas (16) and (17),

$$\overline{y} = \operatorname{argmax}_{y \in Y} U_d(\overline{k}, :) \cdot y \quad (16)$$

$$\overline{k} = \operatorname{argmax}_{k \in M} U_a(k, :) \cdot y \quad (17)$$

In which $U_{tp}\,(k, :)\,(tp = a\ or\ d)$ represents the kth row of the matrix U_{tp}.

Formula (17) reflects the fact that, knowing the defender's strategy y, the attacker would choose the pure strategy $\overline{k}$ which maximizes his own payoff. Formula (16) reflects the fact that, the defender knows the attacker's preference, thus she could also work out the formula (17), getting the $\overline{k}$, based on which she would play the strategy $\overline{y}$ to maximize her own payoff. The defender plays a mixed strategy; therefore, the attacker can only observe the distribution of each pure strategy being played by the defender, but he does not know at an exact time, which pure strategy the defender would play.

In SSE, as defined by formulas (16) and (17), if the attacker is indifferent among several different pure strategies (i. e., he faces a tie), he would choose the strategy which maximizes the defender's payoff (i. e., he breaks the tie preferably to the defender). This is the so-called "breaking tie" assumption in security games. For more information on this assumption, interested readers are referred to von Stengel [32, 33].

For algorithms to find the Strong Stackelberg equilibrium (SSE), interested readers are referred to Conitzer and Sandholm [34].

6.5.4 Bayesian Stackelberg equilibrium

A further extension of the CPP game is the Bayesian Stackelberg CPP game, in which there are multiple types of attackers (i. e., followers), thus the defender (i. e., game leader) can capture the uncertainties of the different attackers (e. g., the terrorists, thieves, activists, etc.). In the Bayesian Stackelberg CPP game, we focus on the uncertainties of the attacker's types (i. e., discrete uncertainties), and not on the continuous uncertainties of the attacker's preferences. Continuous uncertainties are discussed in Section 6.6.

A Bayesian Stackelberg equilibrium (BSE) $(\tilde{k}^1, \tilde{k}^2, \ldots, \tilde{k}^{|\aleph|}, \tilde{y})$ for the Bayesian Stackelberg CPP game is defined by formulas (18) and (19),

$$\tilde{y} = \operatorname{argmax}_{y \in Y} \sum_{l \in \aleph} p^l \cdot U_d^l(\tilde{k}^l, :) \cdot y \tag{18}$$

$$\tilde{k}^l = \operatorname{argmax}_{k \in M^l} U_a^l(k, :) \cdot y \tag{19}$$

In which $\aleph$ is the set of attacker types, such as $\aleph$ = {terrorist, theft, activist}; p^l denotes the threat of the lth attacker, normally represented as probabilities; U_d^l (U_a^l) represents the defender's (attacker's) payoff matrix (see formulas (8) and (9)), when the attacker is the lth type; $M^l = \{1, 2, \ldots, m^l\}$.

For algorithms to find the Bayesian Stackelberg equilibrium (BSE), interested readers are referred to Paruchuri et al. [35].

6.6 Discussion

In the above sections, we briefly introduced the application of game theory to security management in a chemical plant. However, up to now, our discussion has been based on the two strong assumptions mentioned in Section 6.3.1, i. e., A1) rational players; A2) common knowledge of the game. Researchers in the last decade proposed some models and algorithms to deal with security games in which the above two assumptions do not hold.

Figure 6.5 illustrates the uncertainty space in security games, and it is derived from Nguyen et al. [36]. The vertical axis denotes the attacker's payoff uncertainty, which reflects the fact that the defender always faces uncertainties on the attacker's parameters (please also recall the discussion of the parameters of CPP game in Section 6.4.4). The horizontal axis denotes the attacker's rationality uncertainty, which reflects the fact that a rational attacker is not always the case. In Nguyen et al. [36], they also have the third dimension, namely, the defender's strategy uncertainty, which models the fact that the defender might have errors when implementing the optimal mixed strategies. The attacker's observation uncertainty, which models the situation that in security games, the attacker tries to observe the defender's strategy, but there are errors between his observation and the defender's implemented strategy, could also be captured by the defender's strategy uncertainty dimension. In this chapter, for the sake of simplicity, we only show the uncertainty space in two dimensions.

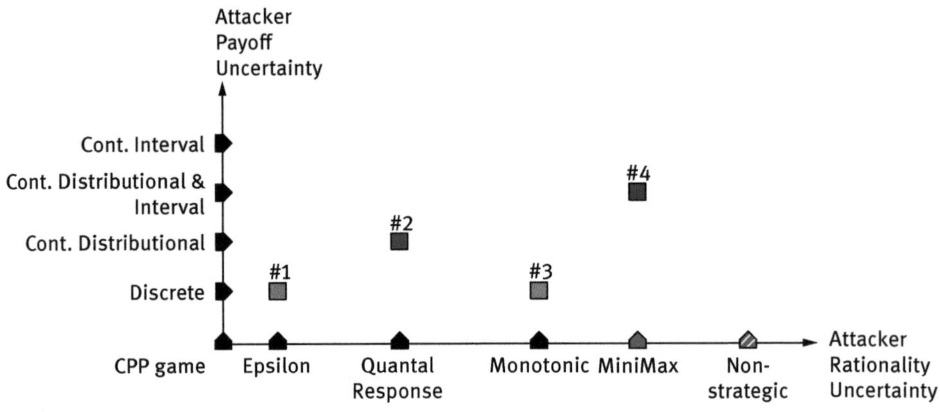

Fig. 6.5: Uncertainty space of CPP game (points are defined in Table 6.2).

As shown in Figure 6.5, the origin of the space is the CPP game, which assumes that both players are rational, and both of them have complete information of the game. The definitions of other points are given in Table 6.2. Some representative references are also given in this table.

Tab. 6.2: Definitions of points in the uncertainty space

Point	Definition
Epsilon	The attacker might have small errors (epsilon) when making decisions, that is, if playing two (or more) strategies would result in very similar payoffs, the attacker might play either. References: Pita et al. [33, 37].
Quantal response	The attacker would not play the best response (a pure strategy which gives the attacker his highest payoff) at probability 1, instead, he would play pure strategies with different probabilities, and these probabilities are calculated by the following formula: $$x_{QRi} = \frac{e^{\lambda \cdot U_a(i,:) \cdot y}}{\sum_{j=1}^{m} e^{\lambda \cdot U_a(j,:) \cdot y}}$$ In which: x_{QRI} represents the probability that the ith pure strategy is played by the attacker; $U_a(i, ;) \cdot y$ denotes the attacker's payoff when the defender plays mixed strategy y while the attacker plays the ith pure strategy; λ is a constant real number, $\lambda = 0$ means the attacker would play each strategy with the same probabilities, $\lambda = \infty$ means that the attacker is a rational attacker. References: McKelvey and Palfrey [38, 39]; Yang et al. [40]; An et al. [41]; Nguyen et al. [42].

Tab. 6.2: (continued) Definitions of points in the uncertainty space

Point	Definition
Monotonic	The attacker would not play the best response (a pure strategy which gives the attacker his highest payoff) at probability 1, instead, he would play pure strategies with different probabilities, and these probabilities satisfy the monotonicity property: $$U_a(i,:)\cdot y > U_a(j,:)\cdot y \Rightarrow x_i > x_j$$ In other words, the monotonic attacker would play a higher payoff strategy with higher probability. Reference: Jiang et al. [43].
MiniMax	The attacker aims at minimizing the defender's maximal payoff, no matter what the defender plays. When the security game is modeled as a zero-sum game, the MiniMax strategy is also the attacker's Nash Equilibrium strategy. This point is emphasized in red color because it is implicitly (at least its idea) used in some non-game-theoretic security risk assessment models, such as in the API SRA framework. Reference: Feng et al. [44].
Non-strategic	The attacker behaves randomly. Evidence shows that attackers would behave according to the defender's defense plan, thus this kind of attacker is not very useful in industrial practice. It is mentioned here acting as a baseline model, and also filling the uncertainty space.
Discrete	Defender has discrete uncertainties on the attacker's parameters. The Bayesian CPP game discussed in section 6.5.4 is a special case of this uncertainty.
Cont. Distributional	Defender has continuous uncertainties on the attacker's parameters. As discussed in section 6.4.4, the defender's estimation of the attacker's estimation on parameters can be a distribution. For example, the defender estimates a target as valuable as € 1M, and she might estimate that a suicide bomber's estimation of this target has a triangle distribution as $C \sim \mathit{Tri}(0.7,1,1.1)$. Reference: Kiekintveld et al. [45].
Cont. Distributional and Interval	Defender has continuous uncertainties on the attacker's parameters. Some parameters are given as distribution, while some of them are given as interval.
Cont. Interval	Defender has continuous uncertainties on the attacker's parameters. And the defender could not even know the distribution of the parameter, however, the defender knows the lower and upper bound of these parameters. For example, if the defender estimates a target as valuable as € 1M, and she might estimate that a suicide bomber's estimation of this target can be as lower as € 0.8M, and as higher as € 1.1M. Any number between 0.8M and 1.1M can be the real parameter of the attacker, but the defender does not know the exact number, and she does not know any further information of this number. References: Kiekintveld et al. [46]; Nikoofal and Zhuang [47]; Zhang et al. [48].
Cross points of Figure 6.5	Different combinations of uncertainties. For instances, the four points #1, #2, #3, and #4 are defined as: #1: defender has discrete uncertainties on attacker's parameters, and the defender thought that the attacker is an epsilon optimal attacker; #2: defender has continuous distributional uncertainties on attacker's parameters, and the defender thought that the attacker is a quantal response attacker; #3: defender has discrete uncertainties on attacker's parameters, and the defender thought that the attacker is a monotonic attacker; #4: defender has continuous interval uncertainties on attacker's parameters, and the defender thought that the attacker is a non-strategic attacker.

Drawing on Figure 6.5, a question occurs, which is whether the space is continuous. Or, in other words, are all points in the space meaningful? For example, what is the meaning of a point between the epsilon and the quantal response? We must emphasize that, to the authors' knowledge, we only consider the points explicitly illustrated in the axis and their cross points (e. g., points like the #1, #2, etc.). Different points in the space have different properties, thus need different models and algorithms. As mentioned in Table 6.2, some points already have corresponding models and algorithms, but some points still need more research effort.

Filling the uncertainty space with models and algorithms, industrial security managers could choose the most suitable point to use according to how many information they have as well as how they think about their security situation, without caring about the complicated mathematical procedures behind it. Game theoretic models could neither improve security without data nor generate quantitative security data, but what they could do is to make the best use of available data [48].

6.7 Conclusion

Security risk assessment without considering adaptive (intelligent) attackers can be considered to be theoretically unreliable and actually fails to optimally allocate limited resources. Game theory, which was born to deal with intelligent interaction among intelligent players, therefore shows great potential in security applications.

In this chapter, game theory and the security game are briefly introduced. An application of game theory for managing security in a chemical plant, which is based on the general intrusion detection approach in a chemical plant, is illustrated. Furthermore, models and algorithms for dealing with uncertainties in security games are introduced and explained.

References

[1] Brown G, Carlyle M, Salmerón J, Wood K. Defending critical infrastructure. Interfaces. 2006;36(6):530–44.

[2] Reniers GLL, Cremer K, Buytaert J. Continuously and simultaneously optimizing an organization's safety and security culture and climate: The improvement diamond for excellence achievement and leadership in safety & security (IDEAL S&S) model. J Clean Prod. 2011;19(11):1239–49.

[3] Bajpai S, Gupta J. Site security for chemical process industries. J Loss Prev Process Ind. 2005;18(4):301–9.

[4] ACS. Security Risk Factor Table. San Diego: Advanced Chemical Safety Inc., 2002.

[5] Security Risk Assessment Methodology for the Petroleum and Petrochemical Industries, (2013).

[6] Cox Jr LAT. Some limitations of "Risk= threat× vulnerability× consequence" for risk analysis of terrorist attacks. Risk Anal. 2008;28(6):1749–61.

[7] Powell R. Defending against terrorist attacks with limited resources. American Political Science Review. 2007;101(03):527–41.
[8] Cox Jr LAT. Game theory and risk analysis. Risk Anal. 2009;29(8):1062–8.
[9] Nisan N, Roughgarden T, Tardos E, Vazirani VV. Algorithmic game theory: Cambridge University Press Cambridge; 2007.
[10] Alpcan T, Başar T. Network security: A decision and game-theoretic approach: Cambridge University Press; 2010.
[11] Pita J, Jain M, Marecki J, Ordóñez F, Portway C, Tambe M, et al., editors. Deployed ARMOR protection: the application of a game theoretic model for security at the Los Angeles International Airport. Proceedings of the 7th international joint conference on Autonomous agents and multiagent systems: industrial track; 2008: International Foundation for Autonomous Agents and Multiagent Systems.
[12] Zhang L, Reniers G. A game-theoretical model to improve process plant protection from terrorist attacks. Risk analysis: An official publication of the Society for Risk Analysis. 2016.
[13] Pavlova Y, Reniers G. A sequential-move game for enhancing safety and security cooperation within chemical clusters. J Hazard Mater. 2011;186(1):401–6.
[14] Talarico L, Reniers G, Sörensen K, Springael J. MISTRAL: A game-theoretical model to allocate security measures in a multi-modal chemical transportation network with adaptive adversaries. Reliab Eng Syst Saf. 2015;138:105–14.
[15] Reniers G, Van Lerberghe P, Van Gulijk C. Security risk assessment and protection in the chemical and process industry. Process Saf Prog. 2015;34(1):72–83.
[16] Rapoport A, Chammah AM. Prisoner's dilemma: A study in conflict and cooperation: University of Michigan press; 1965.
[17] Von Neumann J, Morgenstern O. Theory of games and economic behavior: Princeton university press; 2007.
[18] Nash JF. Equilibrium points in n-person games. Proc Nat Acad Sci USA. 1950;36(1):48–9.
[19] Harsanyi JC. Games with incomplete information played by "Bayesian" players, i–iii: part i. The basic model &. Management science. 2004;50(12_supplement):1804–17.
[20] Tambe M. Security and game theory: algorithms, deployed systems, lessons learned: Cambridge University Press; 2011.
[21] Stackelberg Hv. Theory of the market economy. 1952.
[22] Rao NS, Poole SW, Ma CY, He F, Zhuang J, Yau DK. Defense of Cyber Infrastructures Against Cyber-Physical Attacks Using Game-Theoretic Models. Risk Anal. 2015.
[23] (FAS) FoAS. Al qaeda training manual. 2006.
[24] Zhuang J, Bier VM. Balancing terrorism and natural disasters-defensive strategy with endogenous attacker effort. Operations Research. 2007;55(5):976–91.
[25] Azaiez MN, Bier VM. Optimal resource allocation for security in reliability systems. European Journal of Operational Research. 2007;181(2):773–86.
[26] gametheory.net. Rationality. Available from: http://www.gametheory.net/dictionary/Rationality.html.
[27] Guikema SD. Game theory models of intelligent actors in reliability analysis: An overview of the state of the art. Game theoretic risk analysis of security threats: Springer; 2009. p. 13–31.
[28] Rios J, Insua DR. Adversarial risk analysis for counterterrorism modeling. Risk Anal. 2012;32(5):894–915.
[29] gametheory.net. Mixed strategy. Available from: http://www.gametheory.net/dictionary/MixedStrategy.html.
[30] Gibbons R. A primer in game theory: Harvester Wheatsheaf; 1992.
[31] Lemke CE, Howson J, Joseph T. Equilibrium points of bimatrix games. Journal of the Society for Industrial and Applied Mathematics. 1964;12(2):413–23.
[32] Von Stengel B, Zamir S. Leadership with commitment to mixed strategies. 2004.

[33] Pita J, Jain M, Tambe M, Ordóñez F, Kraus S. Robust solutions to Stackelberg games: Addressing bounded rationality and limited observations in human cognition. Artificial Intelligence. 2010;174(15):1142–71.
[34] Conitzer V, Sandholm T, editors. Computing the optimal strategy to commit to. Proceedings of The 7th ACM Conference on Electronic Commerce; 2006: ACM.
[35] Paruchuri P, Pearce JP, Marecki J, Tambe M, Ordonez F, Kraus S, editors. Playing games for security: an efficient exact algorithm for solving Bayesian Stackelberg games. Proceedings of The 7th International Conference on Autonomous Agents and Multiagent Systems-Volume 2; 2008: International Foundation for Autonomous Agents and Multiagent Systems.
[36] Nguyen TH, Jiang AX, Tambe M, editors. Stop the compartmentalization: Unified robust algorithms for handling uncertainties in security games. Proceedings of The 2014 International Conference on Autonomous Agents and Multiagent Systems; 2014: International Foundation for Autonomous Agents and Multiagent Systems.
[37] Pita J, Jain M, Ordóñez F, Tambe M, Kraus S, Magori-Cohen R, editors. Effective solutions for real-world stackelberg games: When agents must deal with human uncertainties. Proceedings of The 8th International Conference on Autonomous Agents and Multiagent Systems-Volume 1; 2009: International Foundation for Autonomous Agents and Multiagent Systems.
[38] McKelvey RD, Palfrey TR. Quantal response equilibria for extensive form games. Experimental economics. 1998;1(1):9–41.
[39] McKelvey RD, Palfrey TR. Quantal response equilibria for normal form games. 1993.
[40] Yang R, Ordóñez F, Tambe M, editors. Computing optimal strategy against quantal response in security games. Proceedings of the 11th International Conference on Autonomous Agents and Multiagent Systems-Volume 2; 2012: International Foundation for Autonomous Agents and Multiagent Systems.
[41] An B, Ordóñez F, Tambe M, Shieh E, Yang R, Baldwin C, et al. A deployed quantal response-based patrol planning system for the US Coast Guard. Interfaces. 2013;43(5):400–20.
[42] Nguyen TH, Yang R, Azaria A, Kraus S, Tambe M, editors. Analyzing the Effectiveness of Adversary Modeling in Security Games. AAAI; 2013.
[43] Jiang AX, Nguyen TH, Tambe M, Procaccia AD, editors. Monotonic maximin: A robust stackelberg solution against boundedly rational followers. International Conference on Decision and Game Theory for Security; 2013: Springer.
[44] Feng Q, Cai H, Chen Z, Zhao X, Chen Y. Using game theory to optimize allocation of defensive resources to protect multiple chemical facilities in a city against terrorist attacks. J Loss Prev Process Ind. 2016;43:614–28.
[45] Kiekintveld C, Marecki J, Tambe M, editors. Approximation methods for infinite Bayesian Stackelberg games: Modeling distributional payoff uncertainty. The 10th International Conference on Autonomous Agents and Multiagent Systems-Volume 3; 2011: International Foundation for Autonomous Agents and Multiagent Systems.
[46] Kiekintveld C, Islam T, Kreinovich V, editors. Security games with interval uncertainty. Proceedings of The 2013 International Conference on Autonomous Agents and Multiagent Systems-Volume; 2013: International Foundation for Autonomous Agents and Multiagent Systems.
[47] Nikoofal ME, Zhuang J. Robust allocation of a defensive budget considering an attacker's private information. Risk Anal. 2012;32(5):930–43.
[48] Zhang L, Reniers G, Qiu X. Playing chemical plant protection game with distribution-free uncertainties[J]. Reliability Enhineering & System Safety, 2017.
[49] Lathrop J, Ezell B. Validation in the absence of observed events. Risk Anal. 2015.

Nicola Paltrinieri, Cecilia Haskins

7 Dynamic security assessment: benefits and limitations

Abstract: Larger amounts of information are progressively becoming available through new information and communication technologies. This leads to challenges in protecting sensitive systems from nations and critical infrastructures to power and manufacturing plants. Security must apply on a twofold dimension including physical and cyber spaces. Such complexity increases an already large attack surface such that analytical tools assume an essential role not only for the design of appropriate security measures but also for their maintenance and improvement. This study addresses the challenge of security evaluation as a support for decision making using advanced methods and an overall dynamic approach that provides continuous update and refinement of quantitative assessments. This approach allows potential users to understand the probability of both threats to the system and failure of countermeasures. This way priorities for maintenance and improvement of security measures can be defined thoroughly. Dynamic security assessment may lead to improved evaluation and continuous reduction of related uncertainties. However, tight dependence on human behavioral data may challenge private and democratic norms. This suggests formalization of dynamic security assessment considering the benefits and limitations outlined as goals for future research.

7.1 Introduction

Recent advances of information and communication technologies are reshaping our society. Larger amounts of information are potentially available through tighter networks with increased interconnections. This fact may lead to previously inconceivable possibilities such that protecting sensitive systems (from nations and critical infrastructures to power and manufacturing plants) has become increasingly challenging. Security should now be guaranteed on a twofold dimension including the physical and cyber spaces of an asset [1]. Such complexity continuously increases the attack surface and analytical tools assume a fundamental role not only for the design of appropriate security measures but also for their maintenance and improvement. For this reason decision making for security needs an appropriate framework that can handle challenges from current heterogeneous dynamic and interconnected systems.

The objective of this contribution is to present an approach for dynamic security assessment based on a literature review of analogous methods and highlight related benefits and limitations to consider for further research developments. Section 7 pro-

https://doi.org/10.1515/9783110499087-007

vides a definition of security; it suggests potential integration with safety concepts and introduces the need for continuous evaluation. Section 7.3 takes the topic of continuous security evaluation further by describing advanced assessment methods and outlining an overall approach for dynamic assessment. Section 7.4 presents benefits and limitations by referring to the bigger picture of the dynamic risk management framework. Section 7.5 draws overall conclusions from the study.

7.2 Security: integration and continuous evaluation

The Oxford English Dictionary defines security as "the state of being free from danger or threat" [2]. Security may be considered for a wide range of cases depending on the type of system and threat: cloud-enabled internet of controlled things [3] crime and cyber criminal incentives [4] cyber physical systems [5] data leakage [6] electric infrastructures [7] Internet of Things [8] intrusion detection systems [9] social networking and deception [10] and voting systems [11].

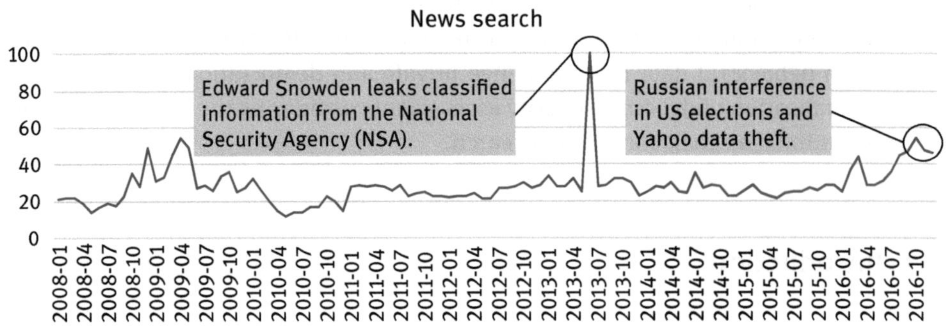

Fig. 7.1: Search popularity of security-related news between 2008 and 2016 as reported by Google Trends [12].

The relative popularity gained by security within the academic sector can be attributed partially to related events reported in the news channels. Google Trends [12] a public web facility of Google Inc. based on Google Search shows how often a particular search term is entered relative to the total search-volume. Figure 7.1 shows the interest in security-related news over time on a quarterly basis. The peak popularity for the term (value of 100) is experienced when Edward Snowden leaked classified information from the National Security Agency (NSA) [13] – in this case the term is also included in the name of the agency. Moreover a recent increase of interest in security corresponds to the news on Russian interference in the US elections [14] and the episodes of data theft from Yahoo [15].

Despite the fact that security and associated information is indisputably a growing trend the results shown by Figure 7.1 deserve further scrutiny. The word and meaning of security is often confused with safety (safety and security are translated with the same word in several languages) but they do not refer to the same concept. Rather the two terms may be considered complementary as stated by Line et al. [16]. Security refers to the case in which a system is free from unwanted influence from the environment in which it is set. Safety conceptually describes the converse: the environment is free from being affected in an undesirable way by the system. The goal of security is to protect physical systems and the information they depend on from malicious parties outside the system boundary. Safety initiatives aim at protecting humans health and the natural environment from damage the system may cause whether intentionally or unintentionally.

Modeling the occurrence of potential unwanted events i. e. their expected frequency and impact supports the definition of appropriate countermeasures for prevention and mitigation of the events. Such measures are often presented as barriers against the development of undesired scenarios. In addition assessing barrier performance increases the understanding of potential system weaknesses and suggests priorities for improvements. Barrier assessments may be carried out for both security and safety purposes but may differ in the priorities identified. For this reason, integration of security and safety assessments potentially supports analysis of both unintentional and intentional events thereby covering both the system and the environment.

However, systems and the environment change with time and the inputs used for barrier performance evaluation may become outdated especially for fast-evolving information and communication technologies. A certain degree of uncertainty may also be associated with the initial information used for such assessment. Generic probabilities of unwanted events may not reflect the actual system/environment conditions. Periodic assessment based on current or more accurate data would address this issue. Dynamic security assessment represents a new frontier for the support of critical decisions related to security strategies because it has the potential to effectively improve the allocation of finances and personnel to the current top-priority counteracting measures.

7.3 Advanced security assessment methods

Sun et al. [17] developed an online dynamic security assessment scheme for large-scale interconnected power systems. In this context security addresses continuous supply in electricity networks. The focus of energy security is often on the risks posed by external threats such as social risks from terrorist or cyberattacks [18] and aims at providing online assessment based on near real time measurements of critical indicators. Indicators for energy security assessment may range from fault variables (fault type and location or outputs of generators) [17] to more sophisticated indexes (such as system average interruption frequency system average interruption duration and customer average interruption duration) [18].

The dynamic method by Sun et al. [17] is based on logic trees and employs weighted summations to aggregate information from indicators. Moreover it focuses on the development of self-adaptive rules shifting from the “black box” concept to the “white box” concept. Such features resemble the structure of the “risk barometer” (RB) methodology which was developed by Paltrinieri and Hokstad [19] to ensure safety within the oil and gas industry. The model followed by RB represents an example of an approach whose focus can be shifted from safety to security with appropriate settings.

RB is based on articulated logic trees such as bow tie diagrams [19]. A bow tie diagram is a flexible representation of a potential unwanted scenario derived from combining the fault tree and event tree associated with a critical (or top) event [20]. The fault tree identifies the possible initiating events whereas the event tree shows the possible outcomes of the scenario [21]. Figure 7.2 illustrates a simplified bow tie diagram for security assessment. The scenario is triggered by intentional events such as crime or cyber crime which can both lead to event consequences that impact the system such as malicious acts or information thefts. The bow tie diagram is centered on the critical event of “breach in”.

Specific physical or nonphysical means planned to prevent control or mitigate such undesired events may be employed as barriers [22]. Guards at the entrance or around the site perimeter (e. g. access control – Figure 7.2) may prevent criminals from physically accessing critical targets which may range from power and manufacturing plants to service companies. Similarly a firewall monitoring and controlling the incoming and outgoing network traffic may stop cyber criminals aiming to breach into the system network. Figure 7.2 illustrates both physical and virtual breaches either of which may lead to malicious acts or information thefts. Criminals can carry out potentially malicious acts such as sabotage terrorism etc. or steal important information from connected storage devices. These actions are also possible by means of a cyber breach of automation cyber-physical systems and cloud computing that offer access to today’s industry and services. In fact, gaining access to the cyber system of a company may allow a criminal to not only access stored data but also control and possibly alter the production parameters of an industrial plant. This could lead to production issues production stops or other potentially dangerous events. Examples of security barriers to mitigate the consequences of a “breach in” after the critical event are encryption of sensitive data and application of material barriers (physical barriers such as fences guardrails containers protective clothing etc. [23]) as shown in Figure 7.2.

The overall results from the application of RB on the bow tie diagram described would be the near real time indication of the risk of malicious acts and information thefts represented by simplified barometers with traffic light scale (Figure 7.2). Lower risk indicates higher security. However, this overall result would not be sufficient for a clear view on how security can be improved or maintained. Therefore RB also adopts a “white box” approach based on drill down capabilities. The high risk of malicious acts shown by the associated barometer (needle between orange and red in the upper right-hand corner) in Figure 7.2 may be due to degraded material barriers with high failure probability (signaled by a red level) which would not be able to mitigate the consequences of a relatively probable breach in (orange level). In turn the high probability of breach in is affected mainly by a firewall with a moderate probability of failure (orange level) that cannot stop highly probable cyber crimes (between orange and red levels). The other crime acts are also rather probable (orange level) but their path would be stopped by efficient access control (low failure probability and green level).

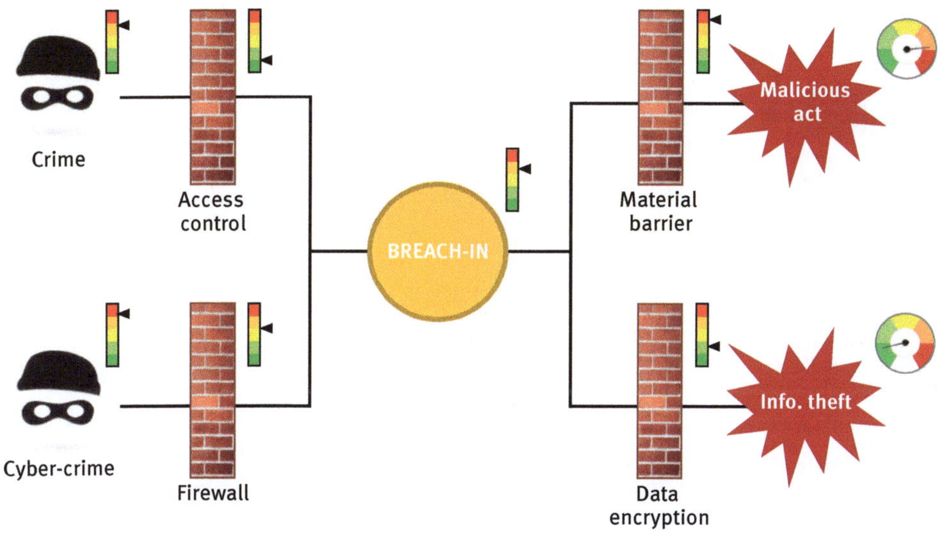

Fig. 7.2: Simplified bowtie diagram for security assessment.

The representative picture of system security assessment shown by Figure 7.2 indicates that there is a high level of (potential) criminality anticipated against which a series of security measures are in place. However, some of these measures (material barriers and firewall) perform poorly and may not be adequate. This suggests the investment in system resources should be focused on the improvement of such measures (e. g. firewall enhancements and material barrier maintenance) in order to decrease the overall risk of malicious acts. Once such measures are improved the security assessment picture is expected to change dramatically. For this reason, RB adopts dynamic assessment with self-adaptive rules. The probability of initial events and the failure probability of security barriers are updated in near real time by means of appropriate sets of indicators [24] (Figure 7.3). These reflect first the probability of breach in and ultimately the risk of malicious acts and information thefts. Dynamic assessment allows for continuously refined results which are representative of the assessed system in the current time rather than relying on generic statistical data derived from other cases.

For instance, several studies focus on crime prediction based specific sets of indicators which may be relatively heterogeneous as shown in Figure 7.3. Bogomolov et al. [25] suggest the use of geo-localized data such as reported criminal cases residential property sales transportation weather and indexes related to homelessness households housing market local government finance and societal wellbeing. Other types of indicators are specifically suitable for continuous collection such as human behavioral data derived from mobile network activity [25] or social network data such as location based network services (Foursquare) [26] or GPS Twitter data [27].

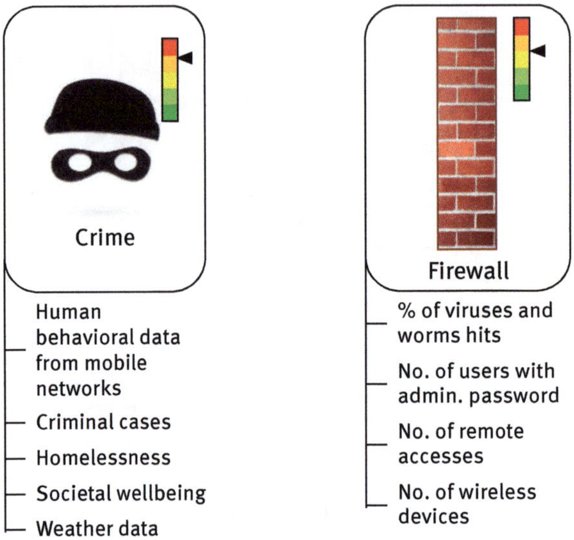

Fig. 7.3: Examples of indicators sets for near real time evaluation of "crime" initiating event and "firewall" security barrier from the bow tie diagram in Figure 7.2.

Torres et al. [28] show how performance of security barriers may be assessed by means of 76 indicators based on specific critical success factors for security effectiveness. Such indicators may directly address the security architecture of the barrier as shown in Figure 7.3 (e. g. percentage of virus and worms hits and number of users with administrator password) or the external and internal connections to the organization intranet or critical data (e. g. number of remote accesses and number of wireless devices). Procedural and organizational indicators are also suggested in order to account for security strategy project accomplishment staff competence commitment and awareness.

Indicators have the potential to improve security decision-making effectively by proactively pointing out increasing threats and degraded barriers. Proactivity is one of main requirements indicated by Johnsen and Øren [29] for the implementation of security measures after the Utøya terror attack (Norway 2011). However, while RB probably would be appropriate for relatively simple scenarios with limited datasets such as the one described in Figure 7.2 advanced models are necessary for interdependent security and interconnected networks. In an interdependent world the risks faced by any system depend not only on its own condition but also on those of others. In the context of terrorism the risks faced by any airline for example are tied to the security standards of other carriers and airports [30]. System complexity leads to large datasets of heterogeneous indicators whose evaluation and process may suffer from challenges of model construction calibration demonstration and reliability. As a result "big data" are changing the landscape of security tools [31] which are increasingly focusing on computational analysis and identification of patterns trends and associations especially relating to human behavior and interactions [32].

System complexity also poses relevant issues for the learning process in the aftermath of unwanted events. While effort is devoted to preventing similar scenarios from repeating the learning process is not isolated from other organizational processes involved in the recovery phase [33] and the lessons learned may be partial irrelevant or simply forgotten. Machine learning and other automatic-learning methods are progressively employed for security assessment in order to both make sense of big data and capitalize on security knowledge from past events [34]. As Morison [35] states online security assessment in which a snapshot of the system condition is captured and analytical engines are used to assess security in near real time has eliminated some of the uncertainty previously introduced by forecasting. New technologies using artificial intelligence and intelligent control systems offer promising results for dynamic security assessment. They may combine the benefits of analytical analysis as well as exhibiting human-like adaptability and decision making capabilities in order to assist system security evaluation and control decisions.

7.4 Benefits and limitations

Dynamic security assessment aims at systematically refining the system security picture based on relevant information and warnings [17 19]. This may lead to a series of benefits. It may fulfil the need for improved quantitative predictions [25–28]. Feedback from continuous assessment iterations may progressively refine generic statistical values [19]. Application of dynamic security assessment not only for design of security measures but also throughout the system lifetime, would allow for precise risk-informed and robust support for security-critical decisions.

Paltrinieri et al. [36] have proposed a framework supporting such an iterative approach: the dynamic risk management framework (DRMF – Figure 4a). DRMF represents an ideal process answering the need for continuous improvement and uncertainty reduction in which dynamic security assessment plays a fundamental role. The framework is composed of four sequential phases and two continuous activities covering the whole process. The process begins with a focus on learning and understanding the potential unwanted scenarios that may occur. Knowledge and information management are addressed by the phases of horizon screening and identification of potential threats. Next the process focuses on decision making. Elaboration and judgement of information and subsequent intervention are addressed by the assessment phase and the decision and action phase. Moving out from the inner zone are two "continuous" activities; namely "communication and consultation" and "monitoring review and continuous improvement." The framework is represented as a shell shaped figure (Figure 7.4a) open to the outside to take into account external inputs effectively during iteration and thereby avoid entrapment in vicious circles.

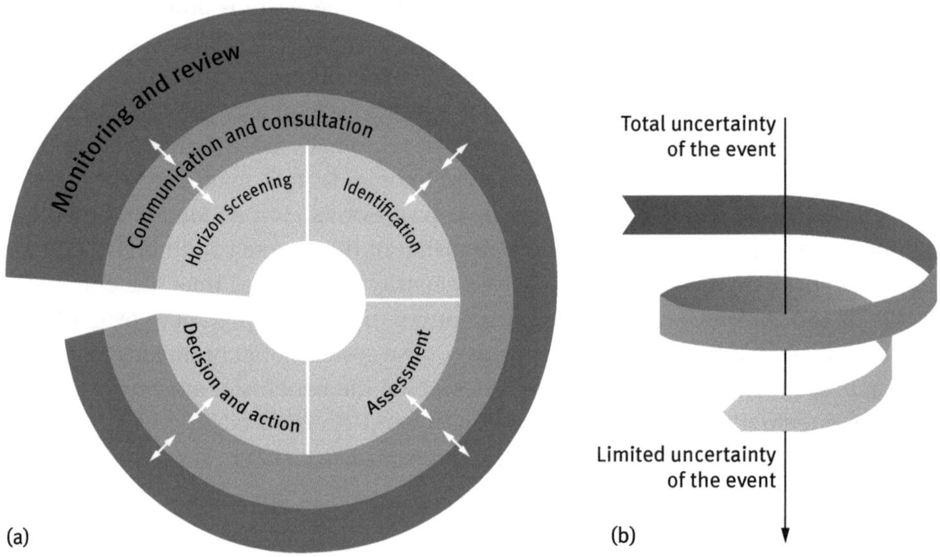

Fig. 7.4: (a) Dynamic risk management framework (adapted from [36]); (b) DRMF iteration towards uncertainty reduction (adapted from [38]).

The main limitations of an iterative approach are its dependence on continuous collection of relevant information along with the need for awareness of appropriate potential threats. In fact there are always potential unwanted events that analysts are not aware that they do not know ("unknown unknowns" as defined by Paltrinieri et al. [37] in relation to risk analysis). The use of dynamic security assessment techniques may lead to increased awareness of previously disregarded threats. The information about potential threats that is continuously collected may raise some reasonable doubts. This information enters the framework through the monitoring of large heterogeneous data sets. Once analysts are aware of potential events that they do not fully understand ("known unknowns" [37]) they can search for and process relevant evidence through the dynamic technique mentioned in this work. As knowledge about these potential unwanted events increases uncertainty decreases (Figure 7.4b). Once new knowledge is metabolized in the process analysts deal with "known knowns" [37] (analysts are aware that they know and may handle such potential events). The dynamic process of risk assessment can be described not only as a circular process (Figure 7.4a) but also as a 3D spiral (Figure 7.4b) where the radial centripetal movement represents the increase of awareness and the vertical movement from the top to the bottom represents the decrease of related uncertainty.

The inputs needed for dynamic security assessment include the aforementioned indicators of the security barrier performance as well as knowledge of human (criminal) behaviors obtained through e. g. mobile network records and social media activity (Figure 7.3). This may promote what Zuboff [39] denounces as an "emergent logic of data accumulation in the networked sphere" leading to implicit surveillance and new expressions of power. Related examples of risk assessment methods based on such human behavioral data have recently emerged [40]:

i) A British firm offering savings on car insurance after assessing the risk of car accidents based on customers' social network profiles.
ii) The Chinese government planning to assess credit risk and assigning "social credit scores" reflecting people's trustworthiness based on various government databases.
iii) A patent by Facebook for assessing credit risk for awarding loans based on Facebook friends' credit scores.

No prior experience prepared people for these new practices. Such unexpected mechanisms of information extraction and control may challenge both privacy and democratic norms [39].

A final limitation of dynamic security assessment is the lack of formalization making this method's application highly case based. To address this challenge expert judgment from potential users may be used preliminarily for the creation and refinement of dynamic security assessment tools. Villa et al. [38] suggest that this mitigates initial skepticism in the model. In the meantime, pre-normative research should commit to define best practices and standards coping with the limitations described.

7.5 Conclusions

This study addresses the challenge of security evaluation as a support for decision making about priorities related to system resource investments. Integration of security and safety assessment should offer a complete risk picture of the system studied and its environment. The study mentions a number of related advanced methods and describes an overall dynamic approach for providing continuous update and refinement of quantitative assessment. This way decision makers can make informed decisions regarding priorities for maintenance and improvement of security measures. In addition, a dynamic approach allows potential users to understand the probability of both threats to the system and failure of counteracting measures. Benefits of dynamic security assessment are improved evaluation and continuous reduction of related uncertainties when associated with appropriate risk awareness. The main limitation is the tight dependence on continuous information whose collection in some extreme cases may challenge private and democratic norms. Such benefits and limitations are important aspects of further research developments and focused formalization of dynamic security assessment approaches.

References

[1] Zhu Q, Alpcan T, Panaousis E, Tambe M, Casey W. Decision and game theory for security. Proceedings 7th International Conference GameSec; Nov 2–4 2016; New York: Springer; 2016.
[2] Stevenson A, editor. Oxford Dictionary of English. Oxford Dictionaries. Oxford; 2016
[3] Cubo J, Nieto A, Pimentel E. A cloud-based Internet of Things platform for ambient assisted living. Sensors. 2014;14(8):14070–14105.
[4] Choo K-KR. The cyber threat landscape: Challenges and future research directions. Comput. Secur. 2011;30(8):719–731.
[5] Cardenas A, Amin S, Sinopoli B, Giani A, Perrig A, Sastry S. Challenges for securing cyber physical systems. Workshop on Future Directions in Cyber-Physical Systems Security 2009.
[6] Shabtai A, Elovici Y, Rokach L. A survey of data leakage detection and prevention solutions. Springer Science & Business Media; 2009.
[7] Sridhar S, Hahn A, Govindarasu M. Cyber–physical system security for the electric power grid. 2012;Proc. IEEE 100(1):210–224. 2012.
[8] Jing Q, Vasilakos AV, Wan J, Lu J, Qiu D. Security of the Internet of Things: perspectives and challenges. Wirel. Networks. 2014;20(8):2481–2501.
[9] Di Pietro R, Mancini LV. Intrusion detection systems Springer Science & Business Media; 2008.
[10] Squicciarini AC, Griffin C. An informed model of personal information release in social networking sites. Privacy Security Risk and Trust (PASSAT) 2012 International Conference on and 2012 International Confernece on Social Computing (SocialCom) IEEE, pp. 636–645.
[11] Gritzalis DA. Secure electronic voting. Springer Science & Business Media; 2012.
[12] Google Inc. Google Trends, Google Search [Online]. Available: www.google.com%5Ctrends. Accessed: Dec 28, 2016.
[13] Greenwald G, MacAskill E, Poitras L. Edward Snowden: The whistleblower behind the NSA surveillance revelations. Guard; 2013.
[14] Entous A, Nakashima E. FBI in agreement with CIA that Russia aimed to help Trump win White House. Washington Post; 2016.
[15] Thielman S. Yahoo hack: 1bn accounts compromised by biggest data breach in history. Guardl; 2016
[16] Line MB, Nordland O, Røstad L, Tøndel IA. Safety vs. Security? Proceedings 8th International Conference on Probabilistic Safety Assessment and Management PSAM. American Society of Mechanical Engineers – ASME; 2006.
[17] Sun K, Likhate S, Vittal V, Kolluri VS, Mandal S. An online dynamic security assessment scheme using phasor measurements and decision trees. IEEE Trans Power Syst. 2007;22(4):1935–1943.
[18] Nepal R, Jamasb T. Security of European electricity systems: Conceptualizing the assessment criteria and core indicators. Int J Crit Infrastruct Prot. 2013;6(3):182–196.

[19] Paltrinieri N, Hokstad P. Dynamic risk assessment: Development of a basic structure Safety and reliability. In Methodology and applications – Proceedings of the European Safety and Reliability Conference ESREL 2014, pp. 1385–1392. Wroclaw Poland: CRC Press; 2015.
[20] Paltrinieri N, Tugnoli A, Cozzani V. Dynamic hazard identification: Tutorial and examples. In Dynamic risk analysis in the chemical and petroleum industry. Butterworth–Heinemann; 2016, pp. 37–48.
[21] Jafarzadeh S, Paltrinieri N, Utne I B, Ellingsen H. "LNG-fuelled fishing vessels: A systems engineering approach. Transp Res Part D Transp Environ. 2017;50:202–222.
[22] Sklet S. Safety barriers: Definition classification and performance. J Loss Prev Process Ind. 2006;19(5):494–506.
[23] Hollnagel E. Barriers and accident prevention. Taylor & Francis; 2016.
[24] Scarponi GE, Paltrinieri N, Khan F, Cozzani V. Reactive and proactive approaches: Tutorials and example. In Dynamic risk analysis in the chemical and petroleum industry. Butterworth–Heinemann; 2016; pp. 75–92.
[25] Bogomolov A, Lepri B, Staiano J, Oliver N, Pianesi F, Pentland A. Once upon a crime: Towards crime prediction from demographics and mobile data. In Proceedings 16th international Conference on Multimodal Interaction. ACM, pp. 427–434; 2014.
[26] Wang M, Gerber MS. Using Twitter for next-place prediction with an application to crime prediction. Computational Intelligence 2015 IEEE Symposium Series on IEEE, pp. 941–948; 2015.
[27] Chen X, Cho Y, Jang SY. 2015 Crime prediction using Twitter sentiment and weather. Systems and Information Engineering Design Symposium (SIEDS) IEEE, pp. 63–68; 2015.
[28] Torres JM, Sarriegi JM, Santos J, Serrano N. Managing information systems security: Critical success factors and indicators to measure effectiveness. In International Conference on Information Security, pp. 530–545. Heidelberg: Springer; 2006.
[29] Johnsen SO, Øren A. 10 years from risk assessment to regulatory action – Is complacency creating a reactive and brittle regulatory regime in Norway? 2015; ESREL.
[30] Heal G, Kearns, M, Kleindorfer P, Kunreuther H. Interdependent security in interconnected networks. In Seeds disaster roots response. How private action can reduce public vulnerability, pp. 258–275. Cambridge University Press; 2006.
[31] Cardenas AA, Manadhata PK, Rajan SP. Big data analytics for security. IEEE Secur Priv 2013;11(6):74–76.
[32] Ward JS, Barker A. Undefined by data: A survey of big data definitions. arXiv Prepr. arXiv1309.5821;2013.
[33] Damlie MA, Nilsen LF, Antonsen S. Learning from disaster – Exploring new ways of seeing. In Safety and reliability of complex engineered systems – Proceedings o25th European Safety and Reliability Conference ESREL, pp. 3731–3738; 2015.
[34] Dua S, Du X. Data mining and machine learning in cybersecurity. CRC Press; 2016.
[35] Morison K. On-line dynamic security assessment using intelligent systems. Power Engineering Society General Meeting 2006, pp. 275–279. IEEE 2006.
[36] Paltrinieri N, Khan F, Amyotte P, Cozzani V. Dynamic approach to risk management: Application to the Hoeganaes metal dust accidents. Process Saf Environ Prot. 2014;92(6):669–679.
[37] Paltrinieri N, Dechy N, Salzano E, Wardman M, Cozzani V. Lessons Learned from Toulouse and Buncefield disasters: From risk analysis failures to the identification of atypical scenarios through a better knowledge management. Risk Anal. 2012;32(8):1404–1419.
[38] Villa V, Paltrinieri N, Khan F, Cozzani V. 2016 Towards dynamic risk analysis: A review of the risk assessment approach and its limitations in the chemical process industry. Saf Sci. 2016;89:77–93.
[39] Zuboff S. Big other: Surveillance capitalism and the prospects of an information civilization. J Inf Technol. 2015;30(1):75–89.
[40] Rutkin A. You are judged on what you post online. New Sci. 2016;232(3099):21.

Nima Khakzad

8 Security vulnerability assessment: A review of Bayesian network approaches

Abstract: In the domain of security risk assessment of critical infrastructures, a number of methodologies has been developed since the 9/11 terrorist attacks in the US. The majority of previous attempts have been devoted to assess the parameters of security risk particularly threat assessment and vulnerability assessment. Among the developed methodologies, the ones based upon Bayesian network, especially when coupled with utility theory and game theory, seem to outperform other techniques. This has been partly due to the capability of Bayesian network in combining objective and subjective information in modeling an uncertain system and partly owing to its ability in accounting for conditional dependencies among the components of the system. In the present chapter, we briefly discuss some previous work which has prominently contributed to the field, followed by some opportunities for further exploiting the features of Bayesian network in security risk assessment.

8.1 Introduction

Factors such as large inventories of hazardous chemicals, which can cause catastrophic consequences if released maliciously, the presence of chemical agents, which can be used either in terrorist attacks or in making chemical and biochemical weapons, and the key role of chemical plants in national and global supply chains have made the security of chemical plants a great concern, especially since 9/11 terrorist attacks in the US. The usage of chemicals in more than half of the terrorist attacks worldwide (Figure 8.1) further emphasizes the security assessment and management of chemical plants.

Terrorist attacks to chemical facilities (even including the ones located in war zones) have been few (Table 8.1). Yet, the risk of terrorist attacks should not be underestimated by authorities and plants' owners and security management; recent attacks to two chemical facilities in France in June and July 2015 raised a red flag about the imminent risk of terrorist attacks to chemical plants in the western world.

The Centre for Chemical and Process Safety (CCPS, 2002) [2] and the American Petroleum Institute (API, 2002) [3] were amongst the first to issue guidelines for security (vulnerability) risk assessment of chemical and petrochemical facilities. In Figure 8.2, the schematic of the methodology developed by API [3] is depicted, where security risk (SR) can be defined as a multiplicative function of threat (T), attractiveness (A), vulnerability (V), and consequences (C) as $SR = T \times A \times V \times C$. By analyzing the threat and the target attractiveness, the attack likelihood can be defined as $L1 = T \times A$, while the success probability of the attack, which is highly dependent on the target vulnerabil-

https://doi.org/10.1515/9783110499087-008

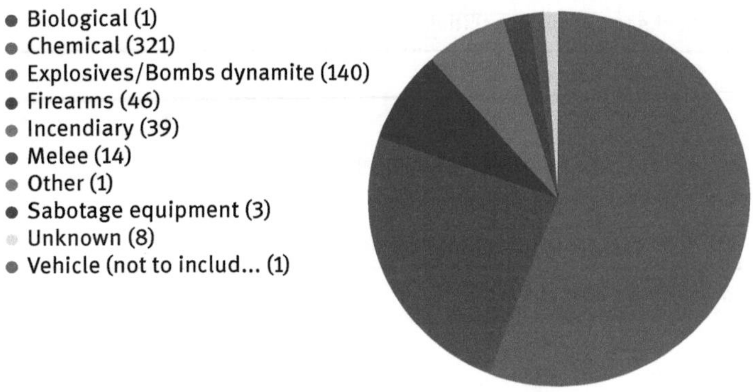

Fig. 8.1: Type of terrorist attacks worldwide in 2015 [1].

ity, is defined as L2 = L1 × V. If the security risk turns out to be unacceptable, it should be reduced by, among other things, allocating and/or upgrading new and/or existing security countermeasures, so that not only the likelihood of a successful attack but also the severity and extent of the consequences can be reduced as SR' = L2' × C'.

Tab. 8.1: Terrorist attacks to chemical facilities

Year	Country	Target
1974	Greece	DOW Chemicals
1983	Peru	Bayer Chemicals
1984	India	Pesticide plant
1985	Belgium	Bayer Chemicals
1990	Libya	Rabta Chemicals
1997	US	Natural gas processing facility
2000	US	Propane storage facility
2001	Yemen	Nexen Chemicals Company
2002	Colombia	Protoquimicos Company
2003	Russia	Storage tanks
2005	Iraq*	Natural gas pipelines
2005	Turkey	Polin polyester factory
2005	Spain	Paint factory
2013	Algeria	Tigantourine Gas Facility
2005	Spain	Metal works facility
2015	France	Chemicals company
2015	France	Storage tanks

* The country was involved in war at the time of attack.

Tab. 8.1 (continued): Terrorist attacks to chemical facilities

Year	Country	Target
2016	Algeria	Krechba Gas Facility
2016	Iraq*	Taji Gas Plant and Energy Facility
2016	Iraq*	Chemical plant
2016	Libya*	Oil storage tank facilities in El-sider

* The country was involved in war at the time of attack.

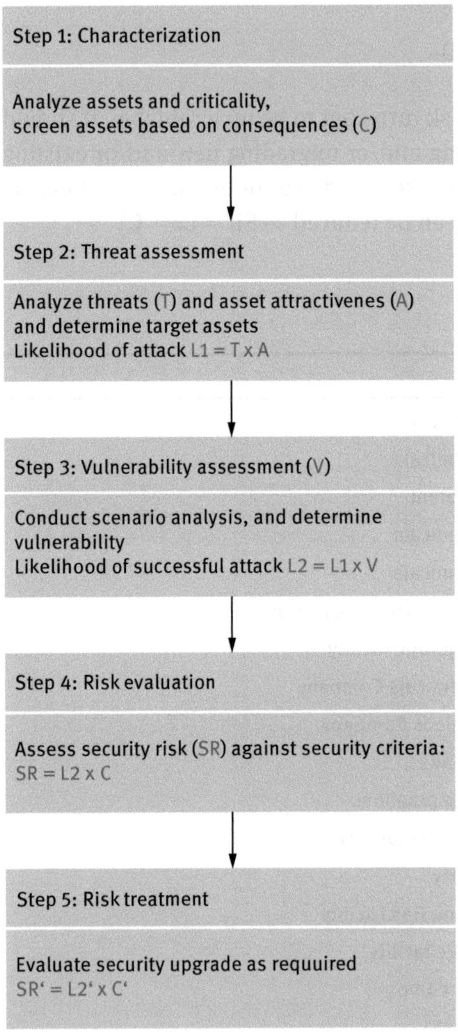

Fig. 8.2: API methodology for security risk assessment [3].

Many techniques and methodologies have been developed that consider the main elements of security risk assessment, that is, threat assessment, vulnerability assessment, modeling attack scenarios, and cost–benefit analysis of countermeasures. Among these the Bayesian network (BN) and its extension and the limited memory influence diagram (LIMID) have been trending due to their flexible structure and robust probabilistic reasoning formalism. Nevertheless, despite the widespread applications of BN and LIMID to safety assessment and risk analysis in the chemical and process industries [4–12], applications to security assessment and vulnerability analysis have been relatively been few [13,14]. Issues such as scarcity of historical data, large uncertainties, subjective probabilities, the need for updating security risks, and complicated conditional dependencies, which all prevail in security risk assessment of chemical and process facilities, however, cannot seem to be addressed appropriately without resorting to BN approaches.

In this chapter, we briefly review previous attempts at applying BN and LIMID to assess security risk and vulnerability, and discuss some new trends in this field while providing grounds for further improvements. Section 8.2 presents the basics of BN and LIMID; applications to the field of security risk assessment are recapitulated in Section 8.3, followed by the further potentiality of BN in addressing the security risk challenges in Section 8.4. The chapter concludes with Section 8.5.

8.2 Bayesian network and influence diagrams

8.2.1 Bayesian networks

BNs [15,16] represent all conditional dependencies among a system's variables by means of joint probability distributions. BNs are acyclic directed graphs in which the systems' random variables (components) are represented by nodes (conventionally, elliptical), while the direct probabilistic dependencies among the nodes are represented by directed arcs. The nodes with arcs directed from them are called parents, while the ones with arcs directed into them are called children. The nodes with no parents are also called root nodes, whereas the nodes with no children are known as leaf nodes (Figure 8.3).

Satisfying the so-called Markov condition, which states that a node (e. g., X_4 in Figure 8.3) is independent of its nondescendants (e. g., X_1 and X_3 in Figure 8.3) given its parents (e. g., X_2 in Figure 8.3), a BN factorizes a joint probability distribution of its random variables (nodes) as a product of the conditional probability distributions of the variables given their parents in the graph:

$$P(X_1, X_2, \ldots, X_n) = \prod_{i=1}^{n} P(X_i | Pa(X_i)) \quad (1)$$

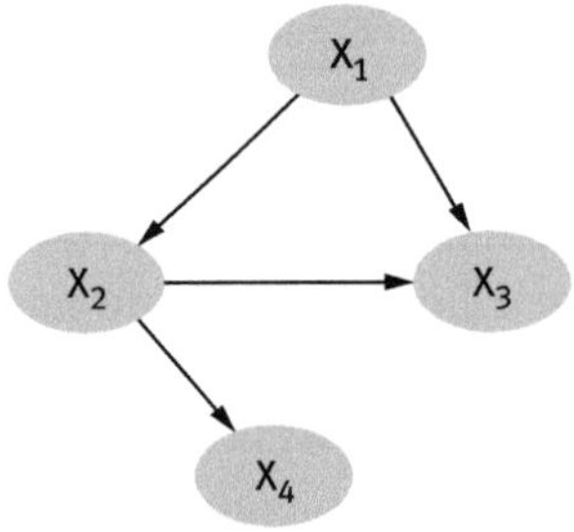

Fig. 8.3: A simple Bayesian network. X1 (root node) is the parent of X2 and X3, and also the ancestor of X4. X2 (intermediate node) is the child of X1, and the parent of X4. X3 (leaf node) is the child of X1 and X2. X4 (leaf node) is the child of X2 and the descendant of X1.

where $Pa(Xi)$ is the parent set of the variable Xi . For example, considering the BN displayed in Figure 8.3, $P(X1, X2, X3, X4) = P(X1)P(X2|X1)P(X3|X1, X2)P(X4|X2)$.

The most important type of reasoning in BNs is probability updating given some information, so-called evidence. The evidence is usually in form of observing a random variable(s) be in one of its states. Accordingly, Bayes' theorem can be employed to propagate the evidence, updating the probabilities of the other nodes conditionally dependent to the observed variable. For example, setting the state of X_4 in Figure 8.3 to one of its states, $X_4 = x_4^+$, the probability of X_1 being in the state x_1^+ can be calculated using:

$$P(x_1^+|x_4^+) = \frac{P(x_1^+)P(x_4^+|x_1^+)}{P(x_4^+)} = \frac{\sum_{X_2,X_3} P(x_1^+,X_2,X_3,x_4^+)}{\sum_{X_1,X_2,X_3} P(X_1,X_2,X_3,x_4^+)} \tag{2}$$

8.2.2 Limited memory influence diagram

BN can be extended to an influence diagram using two additional types of nodes, so-called decision and utility nodes [17]. In order to visually distinguish decision and utility nodes from chance nodes, decision and utility nodes are conventionally displayed in an influence diagram as rectangle and diamond (or hexagon), respectively (Figure 8.4).

Each decision node consists of a finite set of decision alternatives as its states. A decision node should be assigned as the parent of all those chance nodes whose probability distributions depend on at least one of the decision alternatives (e. g., X_2 in Figure 8.4). Likewise, the decision node should be the child of all those chance nodes whose states have to be known to the decision maker before making that specific decision (e. g., X_1 in Figure 8.4).

A utility node is a random variable whose values (utility values) express the preferences of the decision maker as to the outcomes of the decision alternatives. As a

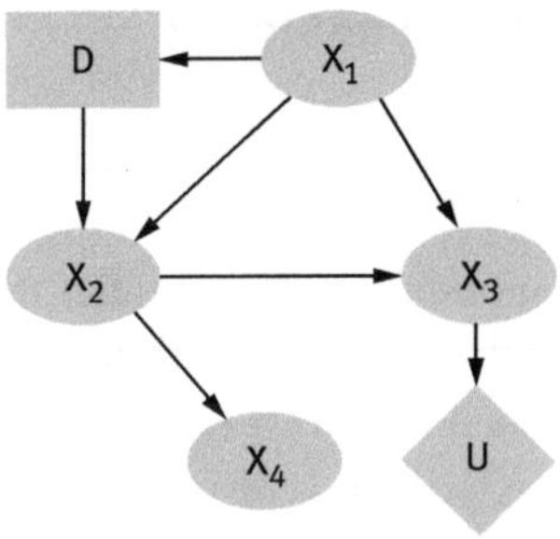

Fig. 8.4: A limited memory influence diagram by adding a decision node D and a utility node U to the Bayesian network.

random variable, each utility node is attributed to a utility table in which the numbers are not probabilities (unlike CPT) but rather utility values (positive or negative) determined by the decision maker for each configuration of parent nodes, either decision nodes or chance nodes.

For example, considering a set of three mutually exclusive decision alternatives for the node $D = \{d_1, d_2, d_3\}$ and two states for the node $X_3 = \{x_3^+, x_3^-\}$ in Figure 8.4, the utility table for the node U includes six utility values for combinations of the decision alternatives and the states. Accordingly, the expected utility of the second decision alternative, $EU(d2)$, can be calculated as:

$$EU(d_2) = \sum_{X_3} P(X_3|d_2)U(d_2, X_3) = P(x_3^+|d_2)U(d_2, x_3^+) + P(x_3^-|d_2)U(d_2, x_3^-) \quad (3)$$

As a result, the decision alternative with the maximum expected utility can be selected as the optimal decision, d^*:

$$d^* = \underset{d_i}{argmax}\, EU(d_i) \qquad \text{for } i = 1, 2, 3 \qquad (4)$$

Utility values are usually determined according to the preferences of the decision maker. Utility values can also be generated using appropriate utility functions. A utility function should express how much the decision maker prefers the outcome y_1 over y_2, considering his attitude toward the decision problem of interest and regarding the existing constraints. In the context of risk based decision making, utility functions are determined based upon the attitude of the decision maker to risk, which can be risk averse, risk neutral, or risk seeking. For a detailed discussion about utility functions, see Gilboa [18].

8.3 Application of Bayesian network to security assessment

8.3.1 Threat assessment

Pate-Cornell and Guikema [13] developed an overarching model based on LIMID to identify terrorist attack scenarios (Figure 8.5). In their approach, considering a specific terrorist group A_j, each attack scenario, At, is represented as a set $At_i = \{T_i, W_i, D_i\}$ comprising (i) T: type of target (e. g., a petroleum storage tank), (ii) W: type of weapon (e. g., a conventional explosive), and (iii) D: means of attack delivery (e. g., by means of drones), denoted in Figure 8.5 as shaded nodes.

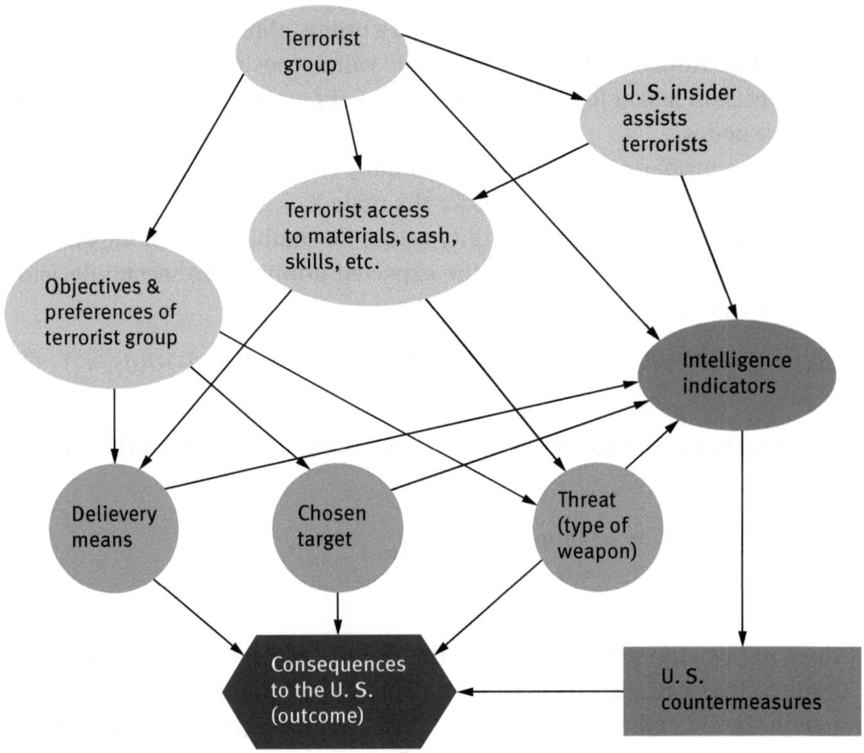

Fig. 8.5: The overarching attack scenario modeling influence diagram [13].

The likelihood of each attack scenario is assessed based on the attackers' capability, intent, and motivation along with the target's attractiveness and vulnerability. The risk of attack scenarios can then be estimated (from the defenders' point of view) as the multiplication of each attack scenario and its expected consequences (denoted as a hexagon in Figure 8.5). Consequences can also be influenced via decisions made by the defender (denoted as a rectangle in Figure 8.5: the US government) as to allocating countermeasures.

Assuming the axioms of rationality [19] – in spite of being violated by human behavior in reality – the expected utility of each attack scenario, $EU_A(At_i)$, based on a specific terrorists' belief about the attack's chances of success and envisaged consequences can be estimated by the defenders. In this step, the defenders put themselves in the shoes of attackers and consider the attack scenario's success likelihood, $P_A(Success|At_i)$, and expected damages, $U_A(Success|At_i)$, from the attackers' perspective as in Equation (5):

$$EU_A(At_i) = P_A(Success|At_i)U_A(Success|At_i) \tag{5}$$

where the subscript A indicates that the analysis is performed by the defenders while they consider the issue from the attackers' point of view. As can be implied by Equation (5), at the time of planning for the attack the attackers are assumed not to be mindful of the defenders' countermeasures, which could otherwise have been reflected in Equation (5) by conditioning the terms on the right-hand side of the equation to the decision alternatives.

The expected utilities computed in Equation (5) for different attack scenarios can be normalized to estimate the probability of each attack scenario given the terrorist group of interest. As such, the conditional probability of the ith attack scenario At_j given the jth terrorist group At_j, that is $P_D(At_i|A_j)$, can be defined by the defenders as:

$$P_D(At_i|A_j) = \frac{EU_A(At_i)}{\sum_{k=1}^{n} EU_A(At_k)} \tag{6}$$

Knowing the general and site specific history of a terrorist group and the intelligence information gathered, the prior probability of the terrorist group $P_D(A_j)$ can be determined by the defender (denoted by subscript D). Considering the availability and performance of the security countermeasures in place, the defenders can also estimate the conditional probability of the success attack given the attack scenario At_j launched by the terrorist group A_j as $P_D(Success|At_i, A_j)$. As such, the success probability of the ith attack scenario by the jth terrorist group can be presented as:

$$P_D(Success, At_i, A_j) = P_D(A_j)P_D(At_i|A_j)P_D(Success|At_i, A_j) \tag{7}$$

As can be noted from Figure 8.5, the defender's countermeasures do not influence the probability of a successful attack but only the envisaged consequences (the only arc from the decision node "US countermeasures" to the utility node "Consequences to the US"); as a result of such a modeling drawback, the last term on the right-hand side of Equation (7), i. e., the conditional probability of success $P_D(Success|At_i, A_j)$, is not conditioned on the US countermeasures (decision alternatives).

Knowing the probabilities of successful attacks for each type of terrorists, the expected utilities can be calculated to prioritize and rank order the security risks via:

$$EU_D(At_i, A_j) = P_D(Success, At_i, A_j)U_D(Success|At_i, A_j, Countermeasures) \quad (8)$$

Based on the same aforementioned priciples, two separate influence diagrams can be used (Figure 8.6) to model the terrorists and the defenders' course of actions based on their respective beliefs accordingly. The left-hand side of Figure 8.6 presents the influence diagram assumed to be used by terrorists as to make decisions about, for example, the potential targets and means of delivery; the right-hand side of Figure 8.6, on the other hand, present the influence diagram used by the defenders to estimate and prioritize security risks and allocate security countermeasures.

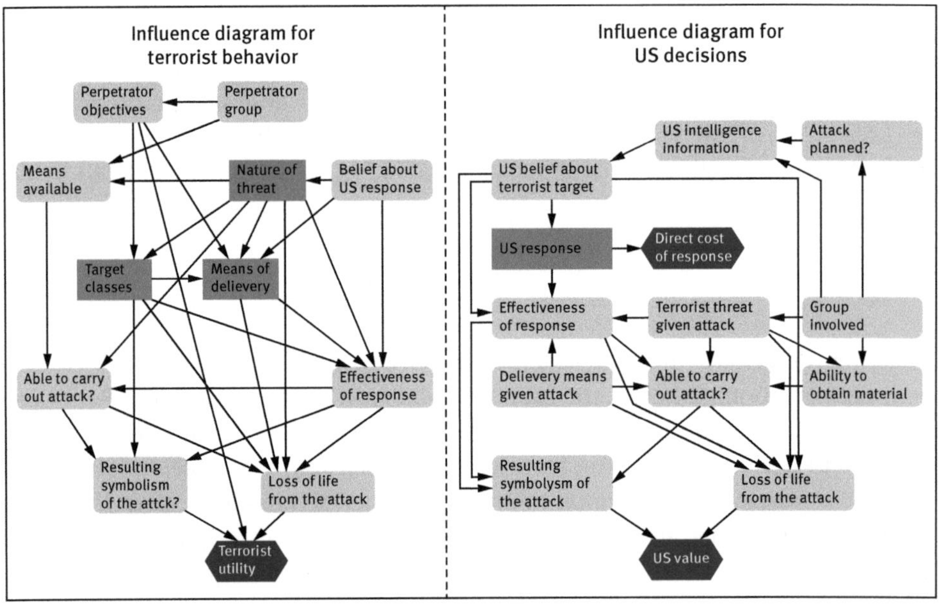

Fig. 8.6: A single-period two-side influence diagram for terrorist risk assessment [13].

A similar approach has been taken in work by Laskey et al. [20], in which two separate BNs were employed simultaneously to detect threatening behavior in the context of cybersecurity. In their work, two identical BNs were developed; one as a generative model to simulate a user's intention and behavior, and the other one as a recognition model to detect threatening user behavior.

As can be noted from the influence diagram in Figure 8.6, the interaction between the terrorists' and defenders' course of actions is very limited since they employ separate influence diagrams for decision making. However, in reality this is hardly the case since the terrorists are able to learn and adjust their attacks according to the countermeasures taken by the defenders and vice versa.

To address this drawback, Rios and Rios Insua [14] proposed several types of game theoretic influence diagrams. Considering a number of defence and attack alternatives as $D = \{d_1, d_2, \ldots, d_m\}$ and $A = \{a_1, a_2, \ldots, a_k\}$, respectively, in a simultaneous defend–attack scenario (Figure 8.7), it is assumed that each decisionmaker assesses separately the probability of a successful attack (S = 1) based on the chosen defence and attack alternatives as $P_D(S = s|d, a)$ and $P_A(S = s|d, a)$. The defender's utility depends on the cost of the countermeasures he chooses and the damage resulting from the attack $U_D(d, s)$, whereas that of the attacker $U_A(a, s)$ reflects the resulting damages:

$$EU_D(d, a) = \sum_{s=0}^{1} P_D(s|d, a) U_D(d, s) \tag{9}$$

$$EU_A(d, a) = \sum_{s=0}^{1} P_A(s|d, a) U_A(a, s) \tag{10}$$

As a result, two different cases can be taken into account.

(i) Common knowledge game: Under the common knowledge assumption, both the attacker and the defender know the expected utility that each pair of (d,a) would provide. As such, a Nash equilibrium (d*,a*) for this game would staisfy:

$$EU_D(d^*, a^*) \geq EU_D(d, a^*) \; \forall d \in D \tag{11}$$

$$EU_A(d^*, a^*) \geq EU_A(d^*, a) \; \forall a \in A \tag{12}$$

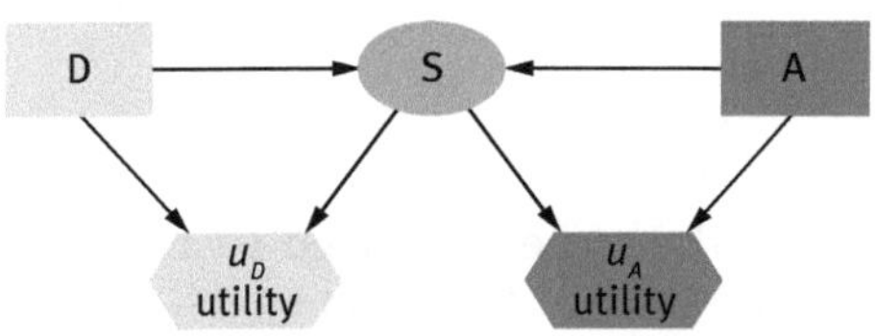

Fig. 8.7: The simultaneous defence–attack influence diagram [14].

(ii) Incomplete knowledge game: Under incomplete knowledge circumstances, each player – whether the defender or the attacker – will be of a certain type, which is known to himself but not his opponent. This way, the players' uncertainties about one another can be presented via prior probability distributions. Figure 8.8(a) depicts a case where the defender chooses a defense alternative $d \in D$ (note the decision node "D" in the figure), the consequences of which (note the utility node "U_D" in the figure) depends on the success of the attack (the node "S") launched by an uncertain type of attack (note the chance node "A"). As can be seen from Figure 8.8(a), the lack of a direct arc from the node "A" to the decision node "D" implies that at the time attack the defender does not know about the type of attacker (i. e., simultaneous defense–attack).

Under such uncertainty, the defender maximizes his expected utility via Equation (13), in which $P_D(A = a)$ presents the defender's uncertainty about the type of attack(er):

$$d^* = \arg\max_{d \in D} \sum_{a \in A} \left(P_D(A = a) \sum_{s=0}^{1} P_D(s|d, a) U_D(d, s)\right) \quad (13)$$

Likewise, the uncertainty of the attacker about the defender type in a simultaneous defense-attack can be modeled using the influence diagram in Figure 8.8(b); subject to a similar uncertainty, the attacker maximizes his expected utility via Equation (14), in which $P_A(D = d)$ represents the attacker's uncertainty about the type of defender:

$$a^* = \arg\max_{a \in A} \sum_{d \in D} \left(P_A(D = d) \sum_{s=0}^{1} P_A(s|d, a) U_A(a, s)\right) \quad (14)$$

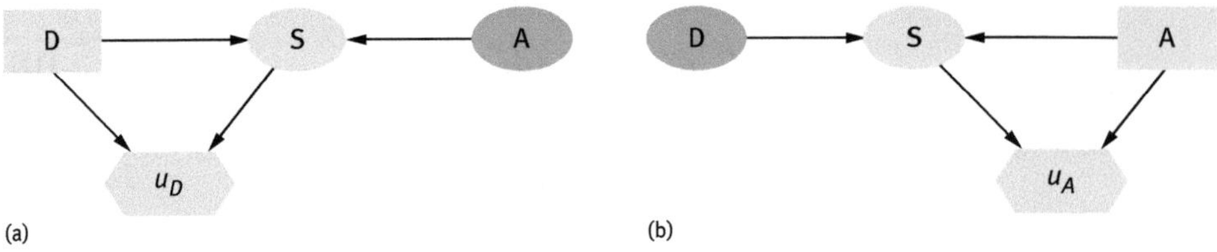

Fig. 8.8: Simultaneous defense–attack model [14]. a) The defender's influence diagram, b) the attacker's influence diagram.

Extending the conceptual models presented in the aforementioned cases, more sophisticated game theoretic influence diagrams can be developed such as sequential defend–attack–defend models (Figure 8.9). As can be noted from Figure 8.9, in a sequential model like this, the attacker launches an attack while having observed the defensive measures taken by the defender (note the arc from the node "D1" to the node "A"); having experienced the attack and knowing the type of attacker, the defender tries to modify his defensive strategy (note the arcs from "D1" and "A" to "D2"). For detailed information the reader is referred to Rios and Rios Insua (2010).

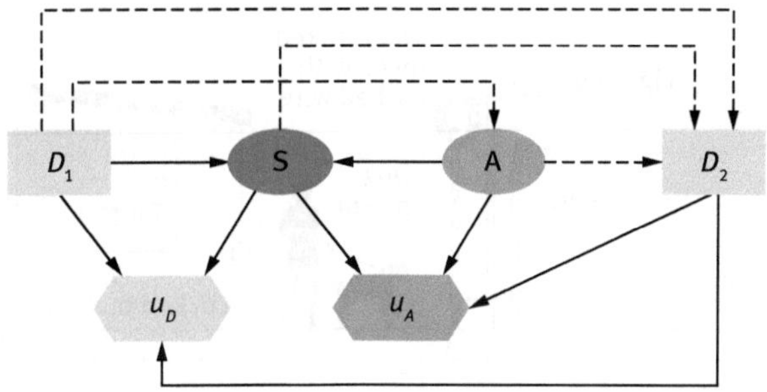

Fig. 8.9: Sequential defense–attack–defense model [14].

8.3.2 Vulnerability assessment

Defined as in the API guideline [3], vulnerability refers to any weaknesses that can be exploited by an adversary to gain access to a target and cause damage. Vulnerability of a target facility depends to a large extent on the availability of security countermeasures and their performance as to deter, deny, detect, delay, and defend – a. k.a 5 D's of outdoor perimeter security – attackers. Having the likelihood of attack determined based upon the threat assessment and target attractiveness analysis, the vulnerability of the target determines the success likelihood of the attack (see Step 3, Figure 8.2).

Garcia [21] introduced a critical path methodology where the probability of a successful attack is estimated based on the time needed by the attacker to disable or penetrate the security measures while undetected. In other words, if the delay time, which is the time required to penetrate a security barrier T_R is less than the guards' response time T_G – a. k.a the mission time – the attacker can successfully penetrate the barrier and reach the target before the arrival of the patrol guards.

Figure 8.10 depicts the schematic of a critical path an attacker may take to the target so as to cause damage, where he should (i) penetrate the wall while not being detected by the motion sensor, (ii) penetrate the fence while not being detected by the camera, and (iii) penetrate the locked building while not being detected by the patrol.

In order to account for the uncertainties embedded in time needed to penetrate a security barrier and the guards' response time, load capacity relationships [23] with random load and capacity viariables can be employed. Accordingly, the probability of successfully penetrating a security barrier can be presented as Equation (15) [24]:

$$P(Success) = P(T_R < T_G) = \int_0^{\infty} \int_0^{T_G} f_R(r) f_G(g) dr dg \tag{15}$$

where T_R is the random variable representing the time required to penetrate the security barrier; T_G is the random variable representing the patrols' arrival time; $f_R(r)$ and $f_G(g)$ are the probability distribution functions for T_R and T_G, respectively, as can be

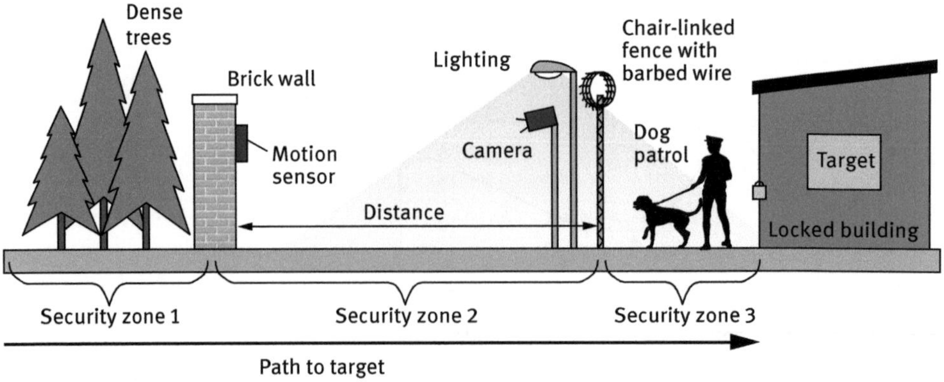

Fig. 8.10: Schematic of the critical path to a target [22].

noted from Equation (15); the longer the patrols' arrival time the higher the probability that the attacker manages to penetrate the security barrier (higher chances of success) [24].

Critical paths are usually protected with more than merely one security barrier; as such, systematic risk assessment techniques are required to model the logical relationships among the available security barriers as to how their failure would help the attacker reach the target.

An attack tree (AT) is such a technique which has widely been used in the vulnerability assessment of cybersecurity systems; yet its application to the domain of physical security has been limited. AT is an excellent tool for brainstorming and evaluating threats while allowing for playing "what –if" games with potential countermeasures. However, similar to the fault tree approach, the AT analysis has a static formalism and is thus unable to account for time dependencies. Although this shortcoming has largely been alleviated through the dynamic attack tree (DAT) technique [25], issues such as common causes and conditional dependencies still cannot easily be handled in DAT. An approach proposed in [25] to solve DAT while considering probabilistic times is illustrated in Figure 8.11.

The dynamic Bayesian network (DBN), on the other hand, is a promising technique with ample application in safety risk analysis and reliability engineering [26–29]. Compared to conventional BN, it makes it possible to account for both spatial and temporal dependencies. These attributes of DBN helps DBN outperform conventional BN in likelihood modeling and safety analysis of stochastic systems. Interval based or discrete time Bayesian network (DTBN) is a form of DBN [26], in which for each random variable the time line $[0, +\infty)$ is partitioned into $n+1$ time intervals, each interval as a state of the random variable (Figure 8.12).

In DTBN, the mission time $[0, T]$ is divided into n intervals (n states) while the last time interval $(T, +\infty)$ is left as the state $n+1$. For example, if random variable X1 in Figure 8.3 is mentioned to be in the jth state $(1 \leq i \leq n)$ or simply $X1 = i$, it means X1

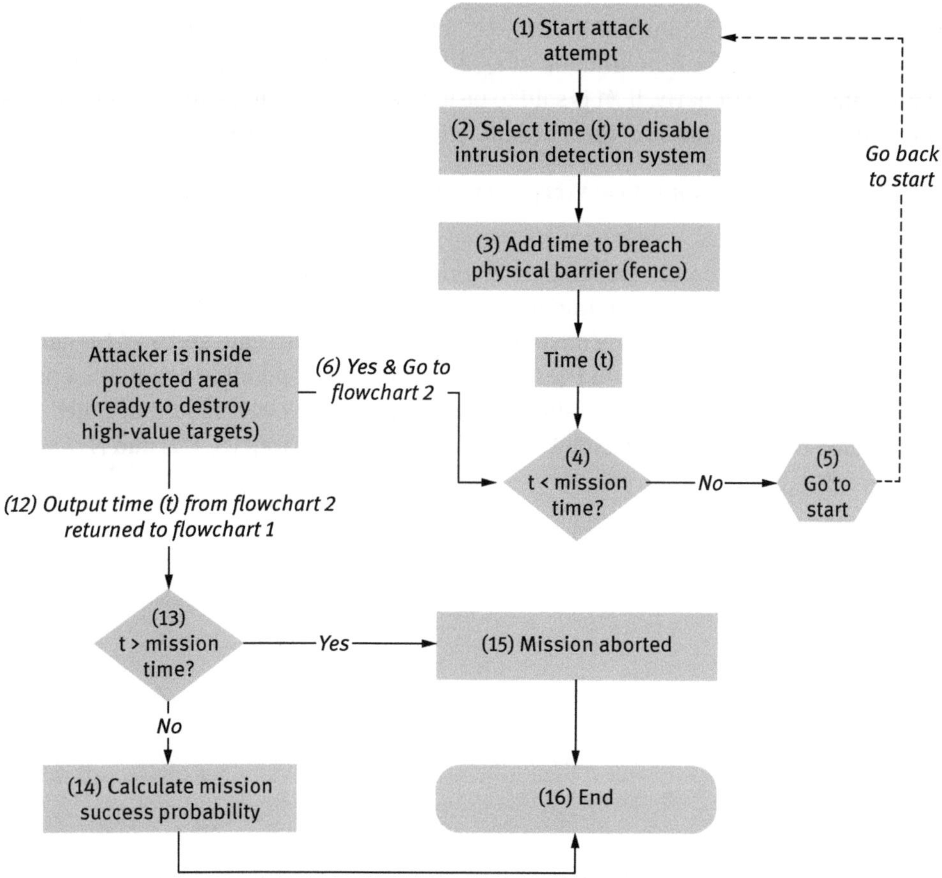

Fig. 8.11: A dynamic approach to solve dynamic attack trees [25].

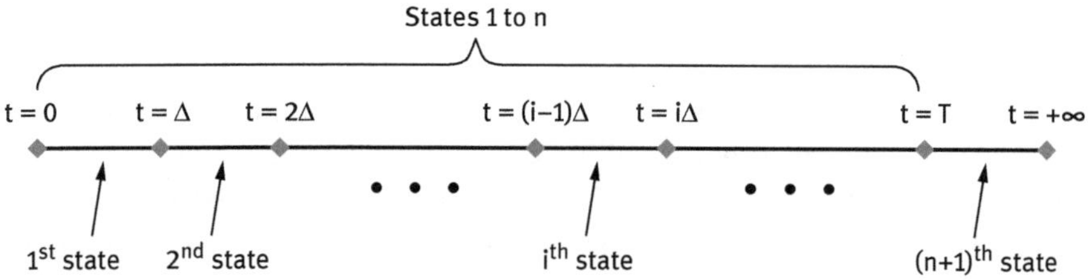

Fig. 8.12: Time intervals as the states of a random variable in discrete-time Bayesian network [29].

has failed in the jth interval (Figure 8.12). In other words, the time to failure of X1 can be shown as $t_{X1} \in ((i-1)\Delta, i\Delta]$. Thus [29]:

$$P(X1 = i) = P((i-1)\Delta < t_{X1} < i\Delta) = \int_{(i-1)\Delta}^{i\Delta} f_{X1}(t)dt = F_{X1}(i\Delta) - F_{X1}((i-1)\Delta) \quad (16)$$

where t_{x1} is the time to failure of X1; $f_{X1}(.)$ is the probability distribution of t_{x1}; $F_{X1}(.)$ is the cumulative distribution function of t_{x1}; Δ is the interval length $\Delta = \frac{T}{n}$, and n is the time granularity. Similarly, if X1 is said to be in the $(n+1)^{th}$ state, which means X1 has not failed during the mission time T (denoted as the survival of X1):

$$P(X1 = n + 1) = P(t_{X1} > T) = \int_T^{\infty} f_{X1}(t)dt = 1 - F_{X1}(T) \quad (17)$$

Using DTBN, the dynamic gates such as priority AND (PAND) gate and sequential failures gate (SEQ) – both take into account the sequential order of failures – can be modeled. For sake of clarity, the modeling of an ordinary AND gate and a PAND gate in the DTBN has been illustrated in Figure 8.13 while the corresponding conditional probability tables have been listed in Tables 8.2 and 8.3, respectively. It should be noted that the mission time T has been divided to equal time intervals, i. e., [0, T/2) and [T/2, T).

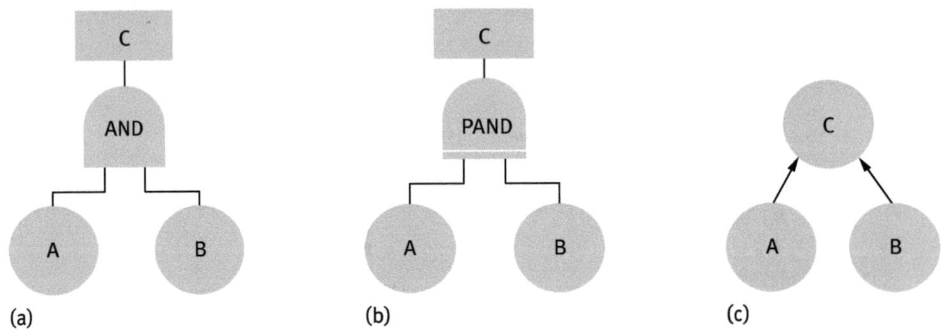

Fig. 8.13: (a) ordinary AND gate: C occurs if and only if both A and B occur, (b) a PAND gate: C occurs if and only if A occurs before B occurs, (c) modeling AND gate and PAND gate in DTBN.

A simplified[1] DAT to assess the vulnerability of the target shown in Figure 8.10 has been displayed in Figure 8.14. Having the probability distribution functions for the times needed to penetrate each security barrier along with the probability distribution of the patrols' response time, the probability of successfully penetrating or disabling each security barrier (the basic events of the DAT) can be calculated in each time interval using Equation (15). The DAT can subsequently be mapped to a corresponding DTBN (Figure 8.15) to account for temporal dependencies and common cause failures (if any). The developed DTBN can especially be used for probability updating and thus updating the target's vulnerability as new information about, among others, the

1 For the sake of simplicity, it has been assumed that the attacker does not need to regress after causing damage to the target.

security barriers' integrity and/or performance become available through monitoring and inspection.

Tab. 8.2: Conditional probability table of an ordinary AND gate in DTBN [27]

A	[0, T/2)			[T/2,T)			[T, ∞)		
B	[0, T/2)	[T/2,T)	[T, ∞)	[0, T/2)	[T/2,T)	[T, ∞)	[0, T/2)	[T/2,T)	[T, ∞)
[0,T/2)	1	0	0	0	0	0	0	0	0
[T/2,T)	0	1	0	1	1	0	0	0	0
[T, ∞)	0	0	1	0	0	1	1	1	1

Tab. 8.3: Conditional probability table of a PAND gate in DTBN [27]

A	[0, T/2)			[T/2,T)			[T, ∞)		
B	[0, T/2)	[T/2,T)	[T, ∞)	[0, T/2)	[T/2,T)	[T, ∞)	[0, T/2)	[T/2,T)	[T, ∞)
[0, T/2)	1	0	0	0	0	0	0	0	0
[T/2,T)	0	1	0	0	1	0	0	0	0
[T, ∞)	0	0	1	1	0	1	1	1	1

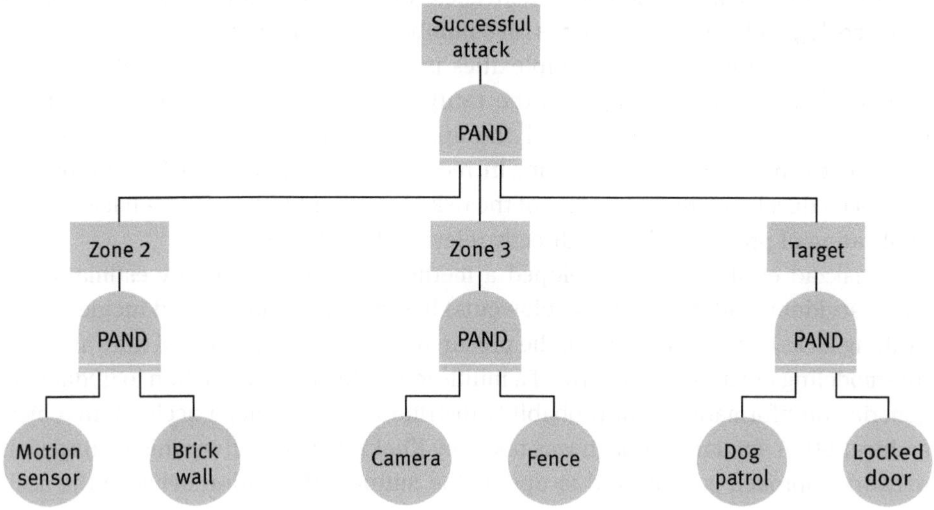

Fig. 8.14: A simplified dynamic attack tree to assess the vulnerability of the target shown in Figure 8.10.

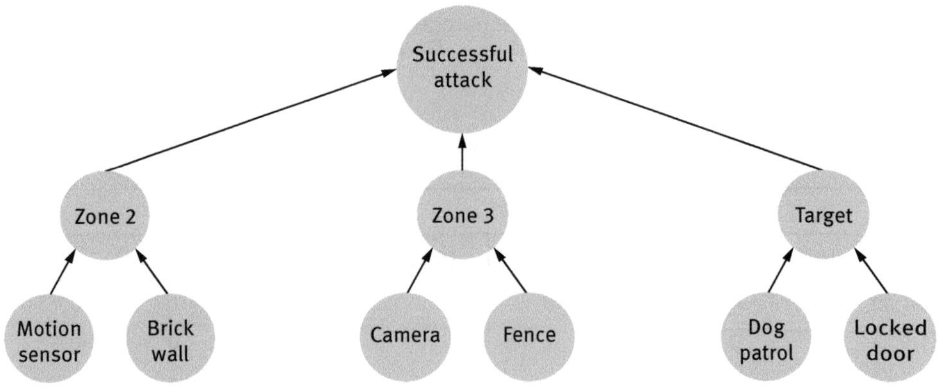

Fig. 8.15: Discrete time BN corresponding to the dynamic attack tree in Figure 8.14.

8.4 Future remarks

8.4.1 Precursor based security assessment

One of the most challenging parameters to estimate in security risk assessment is the attack likelihood. A wide variety of factors should be taken into account in predicting the attack likelihood, including, (i) the general history of threats and attacks to similar targets, locally, regionally, nationally, and internationally, (ii) site specific record of attacks, (iii) capability and potential actions of attackers, (iv) motivation and intent of attackers, and (v) attractiveness of the chemical facility in the eyes of attackers [3], which in turn depends on a number of parameters.

Aside from the foregoing complexities in predicting the attack likelihood, the issue of data scarcity arising from the rarity of terrorist attacks to chemical facilities (Table 8.1) has further limited the application of conventional frequentistic approaches to likelihood estimation. In recent years, a number of BN approaches have been developed in the context of the risk assessment of rare events based on the application of precursor data (indirectly relevant data) [5, 30–32].

Khakzad et al. [30, 31] developed a methodology for probability estimation of major accidents (offshore drilling blowouts) based on near misses and incident data (drilling kicks); in their approach, the precursor data which are elicited from an event tree modeling are used in the form of a multi-nomial likelihood function to update the generic noninformative prior probability distribution of the major accident in a hierarchical BN. A similar approach was taken by Khakzad et al. [32] in which a Bayesian network approach is employed to update the mutual information between the accident precursors and the major accident.

Developing such Bayesian networks for security assessment of chemical plants, a wide variety of security precursor data, including but not limited to the breach of security barriers (even resulting from an unsuccessful attack) or any intentional loss of containment (even not resulting in a major event) can be taken into account to estimate and update the level of the security risk in the plant.

8.4.2 Chemical clusters

The obstacles faced in the security risk assessment of chemical plants seem to be more challenging when it comes to the security assessment and management of chemical clusters. In chemical clusters, besides the complicated interactions among the parameters of security risk assessment (threat, attractiveness, vulnerability, etc.), there lie the structural and infrastructural dependencies and interdependencies, which should be taken into account when assessing and managing the security risks.

Figure 8.16(a) depicts the schematic of a model based on BN that can be used to model the interaction not only among the components of individual chemical plants (the nodes of the same color along with the solid arcs) but also the dependencies among the chemical plants of the cluster (denoted by dashed arcs) [33]. Figure 8.16(b) displays an extension of the same BN as an influence diagram to assess the security risks and expected utilities [33].

Not to mention that in such a BN framework even safety precursor data such as accidental loss of containment, fires, or explosions can be used as indicators to infer the performance and efficiency of barriers, preparedness of employees and emergency response teams, and thus the vulnerability of the plant (and cluster) to intentional events.

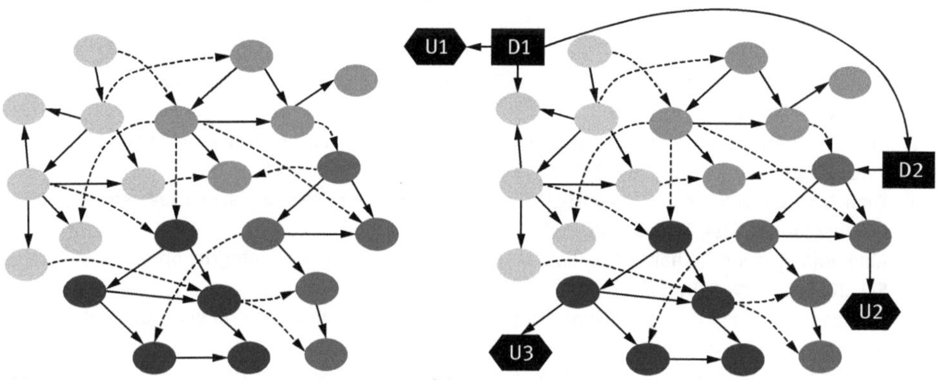

(a) (b)

Fig. 8.16: (a) A BN and (b) an influence diagram to assess security risks in a chemical cluster [33].

8.5 Conclusions

In recent years, a number of methodologies based on the BN approach has been developed to tackle security risks in critical infrastructures. The unique ability of combining objective probabilities with experts' opinions so as to tackle the uncertainties embedded in assessing threats and vulnerabilities 'on the one hand, and accounting for conditional dependencies on the other, have put Bayesian network in the spotlight as a promising technique in security risk assessment.

The foregoing features of Bayesian network when coupled with utility theory and game theory, in the form of influence diagram, have proven efficient in facilitating the modeling of uncertain and complicated interaction between the adversaries – defender and attacker – in a probabilistic framework. Aside from the common practice of BN in threat assessment and vulnerability analysis, interesting applications of (dynamic) BN to precursor based risk assessment and dynamic evolution of stochastic systems – which are a common practice in safety risk analysis – can effectively be investigated in security risk assessment.

References

[1] Global Terrorism Database. Available from: http://www.start.umd.edu/gtd.
[2] Centre of Chemical Process Safety (CCPS). Guidelines for Analyzing and Managing the Security Vulnerabilities of Fixed Chemical Sites. AIChem, ISBN: 978-0-8169-0877-6; 2003.
[3] American Petroleum Institute (API). 2003. Security Vulnerability Assessment Methodology for the Petroleum and Petrochemical Industries. Available from: https://www.nrc.gov/docs/ML0502/ML050260624.pdf.
[4] Khakzad N, Khan F, Amyotte P. Safety analysis in process facilities: Comparison of fault tree and Bayesian network approaches. Reliab Eng Syst Safe. 2011;96(8):925–932.
[5] Khakzad N, Khan F, Amyotte P. Dynamic safety analysis of process systems by mapping bow-tie into Bayesian network. Process Saf Environ. 2013;91(1–2):46–53.
[6] Khakzad N, Khan F, Amyotte P, Cozzani V. Domino effect analysis using Bayesian networks. Risk Analysis. 2013;33(2):292–306.
[7] Khakzad N, Reniers G. Cost-effective allocation of safety measures in chemical plants w. r.t land-use planning. Safety Sci. 2017;97:2–9.
[8] Yuan Z, Khakzad N, Khan F, Amyotte P. Risk analysis of dust explosion scenarios using Bayesian network model. Risk Analysis. 2015;35(2):278–291.
[9] Abimbola M, Khan F, Khakzad N. Risk-based safety analysis of well integrity operations. Safety Sci. 2016;84:149–160.
[10] Khakzad N, Reniers G. Application of Bayesian network and multi-criteria decision analysis to risk-based design of chemical plants. Cheml Eng Trans. 2016;48:223–228.
[11] Khakzad N, Reniers G, Abbassi R, Khan F. Vulnerability analysis of process plants subject to domino effects. Reliab Eng Syst Safe. 2016;154_127–136.
[12] Zarei E, Azadeh A, Khakzad N, Aliabadi M, Mohammadfam I. Dynamic safety assessment of natural gas stations using Bayesian network. Journal of Hazard Mat. 2017;321:830–840.
[13] Pate-Cornell E, Guikema S. Probabilistic modeling of terrorist threats: A system analysis approach to setting priorities among countermeasures. Mil Oper Res. 2002;7(4):5–232.

[14] Rios J, Rios Insua D. Adversarial risk analysis for counterterrorism modeling. Risk Analysis. 2012;32(5):894–915.
[15] Pearl J. Probabilistic reasoning in intelligent systems. San Francisco, CA: Morgan Kaufmann;1988.
[16] Neapolitan R. Learning Bayesian networks. Upper Saddle River, NJ, USA: Prentice Hall, Inc.; 2003.
[17] Jensen FV, Nielsen TD. Bayesian networks and decision graphs. 2nd ed. New York: Springer; 2007.
[18] Gilboa I. Theory of Decision under Uncertainty. New York: Cambridge University Press; 2009.
[19] Von Neumann J, Morgenstern O. Theory of Games and Economic Behavior. Princeton, NJ: Princeton University Press; 1953.
[20] Laskey K, Alghamdi G, Wang X, Barbara D, Shackelford T, Wright E, Fitzgerald J. Detecting threatening behavior using bayesian networks. In Proceedings of the Conference on Behavioral Representation in Modeling and Simulation, May 17, 2004 (Vol. 32, p. 33).
[21] Garcia M. Vulnerability assessment of physical protection systems. Sandia National Laboratories. US: Elsevier; 2006.
[22] McGill W, Ayyub B, Kaminskiy M. Risk analysis for critical asset protection. Risk Analysis. 2007; 27(5):1265–1281.
[23] Ebeling CE. An introduction to reliability and maintainability engineering. 2nd ed. New Delhi: McGraw Hill; 1997.
[24] Fakhravar D, Khakzad N, Reniers G, Cozzani V. Security vulnerability assessment of gas pipelines using discrete-time Bayesian network. Process Saf Environ. 111:714–725.
[25] Khalil Y. A novel probabilistically timed dynamic model for physical security attack scenarios on critical infrastructures. Process Saf Environ. 2016;102:473–484.
[26] Boudali H, Dugan JB. A discrete-time Bayesian network reliability modeling and analysis framework. Reliab Eng Syst Safe. 2005;87:337–49.
[27] Khakzad N, Khan F, Amyotte P. Risk-based design of process systems using discrete-time Bayesian networks. Reliab Eng Syst Safe. 2013;109:5–17.
[28] Khakzad N. Application of dynamic Bayesian network to risk analysis of domino effects in chemical infrastructures. Reliab Eng Syst Safe. 2015;138:263–272.
[29] Khakzad N, Yu H, Paltrinieri N, Khan F. Techniques of reactive frequency update: Bayesian inference. In Khan F, Paltrinieri N (editors.). Dynamic risk analysis in the chemical and process industry. Amsterdam: Elsevier; 2016, pp. 51–61.
[30] Khakzad N, Khan F, Paltrinieri N. On the application of near accident data to risk analysis of major accidents. Reliab Eng Syst Safe. 2014;126:116–125.
[31] Khakzad N, Khakzad S, Khan F. Probabilistic risk assessment of major accidents: application to offshore blowouts in the Gulf of Mexico. Natural Hazards. 2014;74:1759–1771.
[32] Khakzad N, Khan F, Amyotte P. Major accidents (grey swans) likelihood modeling using accident precursors and approximate reasoning. Risk Analysis. 2015;35(7):1336–1347.
[33] Khakzad N, Martinez IS, Kwon HM, Stewart C, Perera R, Reniers G. Security risk assessment and management in chemical plants: challenges and new trends. Process Saf Prog. DOI 10.1002/prs.11914.

Luca Talarico, Genserik Reniers

9 OR methods to enhance security in the chemical process industry

Abstract: This chapter provides a general overview on models and methods borrowed from operations research that can support security decision making. A specific focus on the chemical sector is provided by exploring a wider spectrum of applications aimed at planning for, preventing, responding to, and recovering from disasters triggered by intentional acts such as terrorism. The nature of security-related events in the chemical industry as well the complexity of the associated decision making make the topic very suitable for operations research. In this chapter, operations research applications are classified into four inter-related categories, each one addressing a macro security problem. For each cluster, the ideas behind these OR models are described and the operating principles are briefly summarized. Practical examples and tools are also described. Through a literature review it will be shown how OR could support decision making and enhance the overall security levels in the chemical sector and as a result in the society at large.

9.1 Introduction

Operations research (OR) is a scientific discipline that uses advanced mathematical methods and techniques to support decision making [32]. The Association of European Operational Research Societies describes OR as a scientific approach to solve complex problems. A better understanding of the systems behind these problems will facilitate the choice and the implementation of more effective solutions, which, in general, may involve complex interactions among several factors [19].

As mentioned by INFORM [32], OR encompasses a wide range of problem-solving techniques and methods such as simulation, optimization, queuing theory, probability, statistics, neural networks, expert systems and multi-criteria decision making. Nearly all of these techniques involve the construction of mathematical models that attempt to describe the system. These mathematical models are often characterized either by a maximization problem of one or a combination of measures (e. g., profit, performance, yield) or by a minimization of some indices (e. g. loss, risk, cost). In general, OR supports decision makers in improving the performance of such complex systems by applying algorithms to solve related problems which are depicted by mathematical models [15].

OR methods have been efficiently applied to several domains and in a wide range of sectors such as supply chain, transportation, workforce management, portfolio optimization, project management, education, public utility, health care, telecom-

https://doi.org/10.1515/9783110499087-009

munications and Information technology [61]. Since 9/11 and due to the increasing of terrorism threats, many other areas linked to emergency management, defense and security, have attracted an increased number of researchers and practitioners as the application of OR techniques could significantly enhance the overall levels of security (through efficient and effective planned activities) and thus mitigate the consequences of an intentional act.

Actually, OR had a long and distinguished history in the field of security even before 9/11. For decades, and at least since 1960, the OR community has been exploring and addressing the solutions of security-related problems in the field of border security, port security, airline security and transportation security, etc. [72]. In addition, several authors have reported on the significant impact of science and technology, including OR applications on terrorism response [7, 29]. For example, enhanced decision making supported by OR tools could drastically reduce the vulnerability of critical infrastructure to terrorism, thereby minimizing damage and increasing the ability to recover more effectively from attacks that could potentially occur.

Intentional acts comprise *vandalism, sabotage, terrorism, and cyber attacks*[1] carried out by third parties with the intention of triggering significant consequences. Sabotage and terrorist acts as well as political instability in different countries have clearly increased in the latest decades around the world. These episodes pose significant security threats to the chemical industry because of the increased length and complexity of supply chains [51].

Due to the dangers inherent in the storage, handling, processing, and distribution of hazardous materials, an intentional act targeting a chemical facility might result in a major disaster. In fact, as in the chemical and process industry, dangerous goods are usually stored and/or processed in significant quantities, and intentional act might potentially trigger cascade effects. In addition, due to economies of scale, chemical plants are often grouped in clusters that are geographically close each other, potentially aggravating the destructive consequences of a terrorist attack. Moreover, chemical plants and refineries are often located near major logistic gateways such as ports or highways. In those cases, an attack could potentially disrupt the flow of material and the whole supply chain through those logistic nodes. In other cases, complex process activities are located in densely populated areas, and thus population could obviously be harmed if exposed to chemicals substances and/or hazardous materials [52]. All these elements can pose important security threats to the chemical plan itself as well as to the other plants, infrastructure and/or population located in the nearby. For instance, spills of hazardous materials, fires, explosions and toxic emissions triggered by an intentional accident could affect millions of people with great possibility of mass casualties as well as pro-

1 In the remainder of the chapter if not expressly mentioned there will not be any differentiation between the various types of intentional acts since the potential consequence of an attack, conducted by different parties, will most likely result in the same unwanted consequences.

voking catastrophic damages to the environment. A study conducted by the US Army assessing the biggest US chemical plants in 2001 found that a terrorist attack resulting in a chemical release had the potential to kill or injure about 2.4 million people [36]. In addition, an attack on a chemical plant could have major impact on a nation's economy. In fact, this sector plays an important role in modern society from producing and making available daily necessities such as electronic materials, energy (in the form of power and fuels), medicine and so on. Those goods are essential for our modern way of living [77].

For the above mentioned reasons the chemical and process industry might represent an attractive target for terrorist groups or organizations whose goal is to spread terror and maximize the damage of their actions which can propagate far beyond the direct target, and reverberate long after the immediate damage [44]. In some cases, the risk of a terrorist attack on chemical facilities presents low probability, but nevertheless the consequences of a successful unexpected intentional act can cause huge losses to a nation's economy. For example, repeated attacks targeting refineries and oil pipelines, from 2003 to 2005, in Iraq determined more than $ 10 billion in lost oil revenues [41]. In September 2007, six simultaneous attacks by saboteurs from the people's revolutionary army (EPR) group against oil and gas pipelines caused severe supply shortages in Mexico, leading to the temporary closure of several factories [30]. In June 2016, a terrorist attack on a chemicals industrial plant occurred in France. The attacker drove a car at high speed through the gates of the plant and crashed into gas canisters, causing an explosion. French police discovered a decapitated body inside the car used for the attack. In addition, two other employees of the chemical plant were injured in the attack and the site was locked down [56].

The risks inherent to chemical processes is slowly raising public awareness. Since the attacks in New York City in 2011, greater research efforts from both academia and industry have addressed physical security. In particular, the protection of critical infrastructure and its vulnerability has gained even more attention due to its extreme importance for society and its vulnerability.

The American Department of Homeland Security defines a *critical infrastructure* as "...systems and (physical or virtual) assets so vital for a nation that the incapacity or destruction of such systems and assets would have a debilitating impact on security, national economic security, national public health, or safety..." [8]. The consequences of a terrorist attack for a critical infrastructure might be substantial. For this reason, the protection of industrial process plants, including those in the chemical industry, assume a fundamental importance.

With the threat of terrorism looming, operators and decision makers from the chemical industry have strengthened their actions to prevent terrorists from targeting their infrastructures. However, securing vast chemical plants and/or defining strategies to prevent terrorist attacks and mitigate their consequences might result in a challenging task. Specific features of the security problems are to be evaluated and tight constraints such as limited security budgets are to be considered. In this regard,

OR could be effectively used to enhance decision making in security and risk management related activities by making use of quantitative methodologies.

The goal of this chapter is to describe how OR techniques and tools could help with decisions on security topics, which as a result could increase security in the chemical sector. After having provided a general overview of the methods and algorithms commonly used within OR to support decision making, this chapter focuses on security related challenges that are typical of the chemical sector. Several areas of applications are explored, depending on the security threats and the specific features of the decision making process. For each category practical OR applications and tools used to increase the level of security and to mitigate the consequences of potential attacks are provided.

9.2 General overview on OR methods

OR was initially developed as a scientific discipline during World War II for military applications. At that time, the main goal was to support planning military logistics operations. With the end of the war OR techniques started to be applied also to civil applications for a wide range of problems in several domains. In fact, nowadays OR is effectively employed as a valid decision support system reducing errors and improving the quality within several strategic, tactical and operative decisions. Some nonexhaustive examples of complex problems for which decision makers are widely supported by OR tools include scheduling problems (e. g., production scheduling, staff scheduling, vehicle scheduling), location problems (e. g., definition of the best location for industrial plants, shops, warehouses or strategic facilities), logistics and supply chain problems, forecasting, revenue and pricing problems.

To tackle these complex problems, OR techniques follow a structured approach starting from the mathematical definition of the problem to be addressed. In practice, the complex inter-relations between the main variables that characterize a system are identified. The relationships between the decision variables and the output of the systems are represented in a form of mathematical constraints (e. g., equations and/or inequalities). In addition, one or multiple objective functions are modeled in order to clearly identify the goals that the decision maker intends to achieve.

Mathematical programming, including heuristics, is the most frequently utilized method to solve those models considering either real-life cases or artificial instances. In some cases, due to the inner complexity of the problem, an optimal solution for these instances cannot be achieved. Therefore, by using heuristics, good quality near optimal solutions could be generated for decision making purposes.

As highlighted at Lancaster University Management School [38], the OR techniques, that are most frequently applied to address these problems, are based on the following approaches:

- Simulation: Used to verify the impact of different decisions on solutions. In general, these decisions are tested using pre-built cases and/or real-life scenarios. This way, the impact of different problem parameters on the quality of solutions, as well as the effects produced by different decisions can be "tried out" before being implemented.
- Optimization*:* Generally the only available option, especially when the number of scenarios is so large that a full factorial simulation would be impracticable or even impossible. These approaches are appropriate for complex combinatorial problems dealing with, for example, the allocation of scarce resources, scheduling and routing problems.
- Problem structuring: Often used when dealing with complex decisions that involve many stakeholders with conflicting interests. In those cases, game theoretic models as well as techniques borrowed from decision making theory could be really useful.
- Probability and statistics: Used to analyze data, measure risk, mine data to find hidden patterns and valuable connections, test conclusions and make reliable forecasts.

Focusing on the optimization techniques, on which the majority of the examples are based in the remainder of this chapter, the algorithms that are generally used to solve the decision problems can be grouped in two main categories: exact algorithms and heuristics.

Exact algorithms rely on an exhaustive exploration of the solution space in order to identify the optimal solution. The simplex algorithm, the branch-and-bound method and the column generation algorithm represent some of the most popular exact approaches, which are also embedded in the main commercial solvers available on the market. These methods offer the guarantee of finding optimal solutions to the detriment of long computation time. Depending on the complexity of the problem, exact approaches are subject to a combinatorial explosion as the size of the problem increases. Nevertheless, in many cases, such as when responding to an emergency situation, decision makers need to make a decision in a relatively short time. For these reasons, exact algorithms are not suitable for large and complex decision problems. In those cases, heuristic approaches might be required or recommendable.

Heuristics represent nonexact approaches and are widely employed to solve complex optimization problems. Despite the optimality of the solutions obtained not being guaranteed, heuristics can provide good quality solutions in limited time. Sometimes heuristics are the only viable solution approach, especially when exact methods fail to find any feasible solution due to limiting factors such as time, computation capabilities and so on. In many cases, heuristics are also combined with exact algorithms (hybrid heuristics) to exploit the advantages of both the approaches generating better solutions. Sometimes when dealing with heuristics the term metaheuristics is also used. In practice, a metaheuristic represents a high level problem independent algorithmic framework that provides a set of guidelines to develop heuristic algorithms. The term metaheuristic might also be used to identify a problem specific implementation of a heuristic algorithm according to the guidelines expressed in such a framework [58]. Some metaheuristic approaches include local search based approaches (e. g., iterated local search, GRASP, variable neighborhood search, tabu search) and nature inspired algorithms (e. g., simulated annealing, genetic algorithm, ant-colony optimization).

9.3 OR and security applications

Security management in a dynamic business sector such as the chemical industry requires a multi-disciplinary approach combining a wide range of competencies. As mentioned before, OR could offer a wide portfolio of tools and methods to support decision makers while securing critical infrastructures. Planning activities could be definitely enhanced in order to prevent, respond to and recovery from manmade disaster and terrorism. The challenges and difficulties posed by these security problems (e. g., border and transportation security) have always attracted the attention of researchers in the OR field.

In order to apply traditional OR theories and algorithms to enhance security in the chemical sector a better knowledge and understanding of the features of the underlining security problems might be required. In fact, depending on the security threats and the type of resulting accidents different approaches may be required. Indeed, the effectiveness and the efficiency of existing OR methods can be preserved only if the associated OR models are improved or adapted to respond to specific security related problems, disaster and /or mass emergency situations.

This implies the use of different assumptions while modeling the optimization problem to be addressed. For example, the models behind an optimized emergency/evacuation plan in response to a terrorist act would significantly differ from the approaches to support decision making in the case of accidents triggered by safety-related events. Moreover, when considering terrorist attacks, since intelligent and strategic adversaries are to be faced, traditional methods based on probabilistic risk assessments and historical data analysis cannot always be used.

One of the first attempts to create a framework organizing the scientific literature on OR techniques and methods specifically addressing security issues is due to Wright et al. [72]. In this work, an extensive survey of operations research models and applications in the field of security management is presented.

Tufekci and Wallace [67] analyze the features of the decision models used to support decision making in the case of disaster response (triggered either by a natural or a manmade event) categorizing them into two different classes: *Pre-event models* to prevent and mitigate an unwanted event and *post-event models* to support the response to the event that has occurred. Pre-event models include the ones aimed at predicting and analyzing potential security threats in order to plan the necessary actions for mitigation. Post-event models are used from the moment in time when the disaster begins. The main challenges that a post-event model needs to address are related to the location, the allocation, the coordination and the management of the available resources.

In this chapter, we propose a classification framework of OR applications in the chemical sector made by four different areas: *mitigation*, *preparedness response* and *recovery*. A similar scheme was proposed by Altay and Green [1] while classifying typical activities in the domain of disaster operations management. Following the taxonomy proposed by Tufekci and Wallace [67] the OR applications falling under the mitigation and preparedness clusters can be classified as pre-event models, while the OR methods included in the response and recovery groups represent post-event models. Even if a comprehensive model to coordinate and integrate the mitigation and preparedness planning with the response phase is highly desirable, it is really challenging, or even impossible to develop due to a lack of information and high levels of uncertainty in such situations.

The proposed classification, shown in Figure 9.1, follows the main phases of the disaster life cycle. The categories are strictly correlated and inter-related, since a decision made in an earlier stage could significantly affect the success and the effect of a possible attack. In addition, from a decision model's perspective, the output of a model applied in the mitigation phase could affect the parameters and the variables of a model in the response phase.

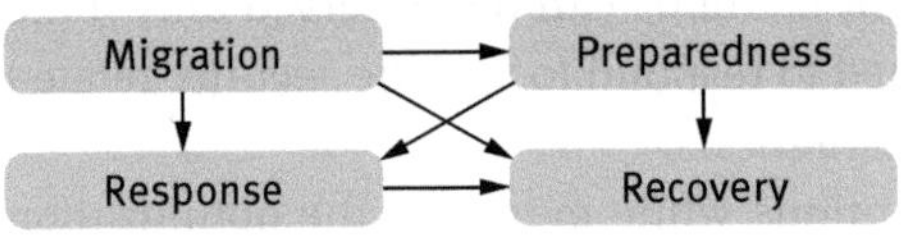

Fig. 9.1: Areas of OR applications.

The category of *mitigation* includes all OR applications aimed at preventing the onset of a disaster due to an intentional act or eventually at reducing the impacts of an attack. The OR models falling in this class focus on strategic and long term decisions including risk analysis, systems design and resource allocation. The OR applications categorized under the umbrella of *preparedness* involve all types of activities aiming at preparing the community to better respond when an attack occurs. *Response* concerns the use of OR methods and tools to develop plans to employ scarce resources and emergency procedures. The final goal is to limit the propagation of an attack, minimize the losses of human life and minimize the damages to property, the environment as well as the social, economic and political structure of a community. As was mentioned before, response activities occur immediately after an attack including the stabilization of the affected areas, providing immediate medical care and evacuation. The OR applications under the label of *recovery* encompass the methods having as main objective the long term recovery after an intentional act. The decisions supported by OR methods focus on repairing the damage generated by the attack to resume the pre-event status. Some applications include the restoration of critical infrastructures, the aid to affected persons, and the coordination of the relief efforts. In general, all OR methods that are grouped in the proposed categories could offer significant contributions in order to improve the decision making process before, during and after an intentionally caused accident. The ultimate goal is to prevent loss of human life, to reduce the negative consequences of the attack on a nation's economy and foster the return to a state of normalcy after the unwanted event.

An alternative way to envision the potential contributions of OR methods falling in the previously mentioned categories is presented in Figure 9.2, which has been adapted from a schema originally proposed by Sheffi and Rice Jr. [57]. The graph shows the theoretical performance evolution of a critical supply chain/infrastructure targeted by an intentional attack. To have an idea of the consequences of a supply chain disruption due to a terrorist attack, Zamparini [76] estimated that the cost on the entire supply chain of a weapon of mass destruction shipped via containers could easily reach 1 trillion dollars, while the direct and indirect costs of the September 11 attacks on the two World Trade Center buildings was estimated at 83 billion dollars.

The proposed groups of OR applications are clearly indicated in the graph in Figure 9.2 to help the visualization of the areas in which decision making could be improved (before, during and after the intentional act). In addition, looking at the vertical axis of the graph, the reader can have a high level indication of the positive contributions that a decision making supported by OR could provide in restoring initial supply chain/infrastructure performance. It should be mentioned that the efficiency of a supply chain could be effectively restored in a relatively short time, minimizing the negative effects of a terrorist attack when, in a planning stage, a supply chain is built considering the positive effect of supply chain redundancy and innovative easy-to-rebuild infrastructure.

In other words, the OR methods included in the proposed four categories (mitigation, preparedness response and recovery) could be used all along the lifecycle of possible intentional act management evolving from a simple security threat into a successful attack, triggering short and long term consequences. As one should expect, each category focuses on a different type of problem for which specific decisions are needed to avoid, contrast and mitigate the consequences of an intentional act.

In the following paragraphs some example of OR optimization methods that encompass possible security applications in the chemical sector are described. The goal is only to provide a high level description of the potentialities of OR in the chemical industry. In fact, this chapter by no means represents an exhaustive bibliography of OR methods in the field of security for chemical processes.

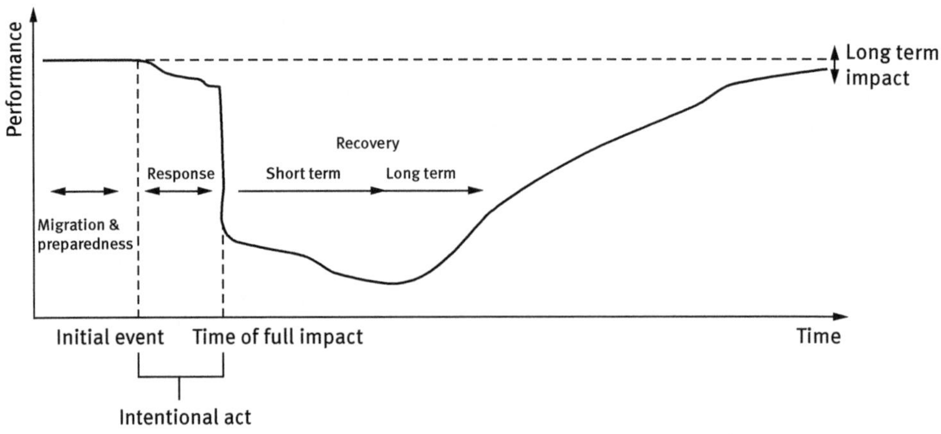

Fig. 9.2: Performance evolution of a critical supply chain/infrastructure affected by a major accident.

9.4 Mitigation

As mentioned before, OR could support decision making during the so-called mitigation phase in which decisions preventing intentional attacks are considered. Those decisions are often related to the location of critical facilities and the levels of security investments to reduce the vulnerability of critical infrastructure and/or to mitigate the unwanted consequence of an attack. Those decisions involve complex infrastructures in which several variables and high levels of uncertainty are to be considered. Quantitative cost–benefit analyses could support such complex security decisions.

In order to better understand the features of the security problem, threat analysis and risk assessment are often performed in a preliminary stage. Traditional methodologies used to perform risk assessment are often based on probabilistic risk assessment using fault and event trees; vulnerability assessment and decision analysis. These methods are, in general, implemented with simulation software and the outcomes are used by risk experts. In order to face the new security challenges posed by international terrorism, traditional probabilistic assessment methods might need to be revised, considering specific features of those intentional attacks, such as the fact that terrorists, in general, carefully select their target to maximize the potential consequence of their actions. Grossi [28] proposes the use of worst case analysis and contingent response as a valid alternative for risk assessment. Srivastava and Gupta [59], for example, developed a methodology for threat analysis and security risk assessment that is specifically adapted to the oil and gas industry.

Focusing on the chemical sector, Fairley [20] discusses some interesting considerations on catastrophic risks. The appropriateness and inappropriateness of different types of risk assessment models for technologically based industrial operations are discussed. Reniers and Dullaert [53] provide an extensive overview of risk assessment methodologies considering security-related attacks on pipeline networks. Reniers et al. [53] study the typology of networks representing industrial areas carefully. A multi-attribute method is developed that models chemical industrial areas as inter-connected and complex networks. A holistic optimization approach is developed considering interorganizational and inter-cluster objectives. The outcome of the model is used to formulate recommendations within a quantitative risk assessment to increase the awareness of decision makers on existing systemic risks. The selection of the most appropriate security measures is not tackled by the proposed optimization method.

Other risk assessment methodologies have been proposed in the OR literature specifically covering routing problems involving the transportation of chemical substances. For example, in Gopalan et al. [26] an attempt to create equity sharing the risk of consequences resulting from an incident among several zones is described. An optimization problem is formulated in which the objective is to establish a set of routes minimizing the total risk of travel and spreading the risk equitably among the geographical regions crossed by vehicles along different routes while transporting hazardous materials. The problem is mathematically stated using an integer programming formulation, and a heuristic solution approach is developed and tested on a real-life routing scenario in a district of New York State.

Assuming that a preliminary risk assessment has been performed in a preliminary phase and the cost effectiveness of available security measures is known in advance, OR can be effectively used to optimize the selection allocation of security resources by solving a knapsack problem with additional constraints such as security budget and or technical limitations. In Talarico et al. [65] the reader can find a practical application of the knapsack problem to secure an illustrative pipeline infrastructure used to transport oil. In Janssens et al.[33], a decision model to reduce the risk of a network breakdown is proposed. Decisions makers in the sector of public utilities are supported by a model that allocates a limited security budget by selecting the most appropriate security measures to increase network security and to minimize the risk of an interruption of service (or material flow) due to external malicious attacks. An effective heuristic algorithm is proposed and tested on a realistic utility network.

In Viduto et al. [69] the security of a telecommunication network is analyzed from a quantitative point of view. Knowing the potential security threats as well as the system vulnerabilities, a security measure selection problem is presented in which both the cost and effectiveness of the available security measures are addressed. A multi-objective tabu search algorithm is developed to construct a set of nondominated solutions. The work proposed by Jerman-Blažič et al. [34] focuses on the prevention of heavy losses due to cyber attacks targeting an ICT network. The authors describe some methods for the identification of cyber security threats and then several decision models, based on cost–benefit analysis, are proposed to select the optimal level of investments in security technology.

In a planning stage, mitigation strategies could also be supported by important strategic decisions on locating critical facilities. This could significantly reduce the risk of a terrorist attack and/or eventually it could lower the negative consequence of an attack, guaranteeing at the same time business continuity and avoiding major disruptions. As also stated by Sheffi and Rice Jr. [57], the ability to quickly recover from major supply chain disruptions can be improved by building redundancy and flexibility into the network. Facility location problems have received great attention in the OR community due to the increasing complexity of these problems and the sometimes conflicting objectives to the optimized. In Current et al. [14] a detailed literature review on facility location problems is presented, which focuses on multi-objective aspects of this research area. In general, these objectives cover four main goals: 1) cost minimization, including distance minimization; 2) demand coverage and demand assignment; 3) profit maximization; and 4) security maximization while dealing with environmental concerns.

As mentioned in the previous paragraph, chemical industrial areas represent critical infrastructures, which due to the type of hazardous materials processed might be subject to a cascade effect. Therefore, in a planning stage, decisions on the layout of chemical installations and on the number and the location of security measures in place might have a significant impact on the likelihood and on the success of a malicious act on such installations. In Reniers et al. [54] some methods are proposed to make a chemical industrial area more "secure by design," attenuating the consequences of intentional attacks. Using a network representation to mathematically represent chemical industrial areas it is proved that the resilience to disaster of such areas may follow a power law distribution. A simulation is carried out to assess the impact on the network when highly hazardous installations are protected against malicious acts. In practice, the network disintegrates into smaller and separate clusters with no escalation danger in between. This way it is possible to protect chemical industrial areas in such a way that they are more resilient against terrorism.

A recent research field that has proven its effectiveness in applications related to security, starting from military applications, over political science to cyber crime, is game theory. Since 9/11, game theory has been used as a tool to support several decisions, such as the selection of security countermeasures, the identification and prioritization of critical assets to defend, the security budget identification and so on [12]. By combining game theoretic models and risk assessment methods, Zhang and Reniers [77] propose a model to improve security in process plants by enhancing their intrusion detection system. Wein and Baveja [70] used a game theoretic model to detect the intrusion of potential terrorist and to launch early warning within a border protection problem. The authors show that the images' quality play an important role in increasing the detection probability and thus the system's performance. Talarico et al. [63] developed a decision model based on game theory that can be used to detect impending attacks on transportation infrastructure and, subsequently, allocate security countermeasures to prevent terrorist attacks. The model focuses on a chemical supply chain that is characterized by the use of different transport modes, each having their own security features, which could be a viable target for terrorist acts.

In the OR community, many models and methods have been developed in the past years that focus on the planning of safe transportation routes, directly or indirectly considering the risk of accidents during transport of chemical and hazardous materials. In the chemical and process industry, routing problems represent a very challenging task where the goal is to minimize population exposure in the event of accidents and/or an intentional attack. OR models to support such type of decisions require a definition of a risk function in order to generate and select routes aiming at the minimization of the operating costs as well as the minimization of risk. In fact, an accident involving the release of hazardous materials might produce severe consequences on the nearby population. Therefore, the risk function is often linked to the possible consequence of the accident and might depend on several factors such as the features of the substance being transported, the road characteristics (e. g., tunnels, road condition, light, traffic, lane width, number of lanes), weather conditions, the population living near the accident and so on [9, 25, 60, 68].

Intentional attacks targeting trucks transporting dangerous substances and/or hazardous materials could also be avoided by increasing the route unpredictability. By generating a set of unpredictable routes, decision makers can easily change their plans to deal with unforeseen circumstances, and, as a result, increase security by making vehicle routes more unpredictable. In this case, decision makers could refer to the so-called "peripatetic" routing models in which nodes are visited several times within a planning horizon, but the use of the same arc twice is explicitly forbidden [10, 46] or rather penalized [65]. This way, it is possible to generate a wide variety of solutions, as required for security reasons. For more details on OR models that can be used to develop optimized routing and increase security levels during transport of hazardous materials, the reader is referred to the survey of Batta and Kwon [4].

9.4.1 Preparedness

Readiness for disasters triggered by intentional acts could significantly be improved if emergency activities to mitigate and limit the consequences of an attack are planned in advance. Some OR works focus on the optimal location of scarce resources such as ambulances and or emergency vehicles/facilities used, e. g., by the fire brigade and rescue teams, in order to maximize the covered areas and minimize the response time. In Farahani et al. [21] a survey on some of the OR locations models and algorithms developed in this field of research are presented.

The problem of locating ambulances requires the definition of deployment sites for emergency vehicles within a specific geographical areas such that response time to reach the potential emergency sites within this area is minimized. Bell et al. [5] proposed a location problem for the selection of emergency aircraft sites to be used in case of security alert for the surveillance and protection of important national areas of interest. A two-step solution approach is developed. In a fist stage, the minimum number of sites is identified using a set covering problem and then the result is improved by finding the p-median solution to the problem ensuring equitable response to all areas of interest. A simulation is also performed to determine the impact of altering aircraft launch and flying times on the number of alert sites required and the amount of coverage provided by selecting fewer locations. Some implications on potential military base closure decisions are also discussed, as well as the tradeoffs between costs and required response times of aircrafts in the case of emergency. Jia et al. [35] analyze existing OR models addressing common emergency situations, such as house fires and regular health care needs. In the second part of their paper, the characteristics of large-scale emergencies resulting, e. g., from terrorist attacks and their impact on locating facilities are discussed. Based on these specific features, a covering model is proposed in order to respond to large-scale emergencies, reducing both loss of life and economic losses. A simulation is performed to optimize the locations of facilities for medical supplies to address large-scale emergencies in Los Angeles.

Other OR applications that can be classified under the category of preparedness concern other types of emergency planning activities. For example, in Bastien et al. [3] the evacuation that needs to be performed in the case of disasters is used as an index to develop a risk/safety-assessment model. A model is proposed to simulate an evacuation plan in the case of nuclear plant accidents. In Scanlon and Cantilli [55] the risk exposure is considered by evaluating the consequences of an incident for the community traversed by truck routes transporting hazardous materials. Those indices are used within a simulation model that consider a road transportation mode and aims at measuring the community preparedness in the case of emergency.

Preparedness for emergency could be significantly improved if a better understanding of the security threat and the potential consequences of an attack are known in advance. This could improve public awareness of risks, driving the requirements for more effective and efficient training activities and disaster exercises for emergency teams. Brown et al. [8] applied simulation to study the impacts of disruptions on critical infrastructure and to evaluate interdependencies with other systems. This way, more informed decisions could be made by decision makers in the case of emergency. Glickman and Rosenfield [24] formulated models to evaluate the risks associated with train derailments and the release of hazardous materials, issues that could become important in the event of a terrorist attack. Larson et al. [39] reviewed the literature on police, fire, and emergency medical services and provided precious recommendations that are useful to respond to an emergency in the case of hazardous materials and bioterrorism. Another interesting survey on emergency response is presented by Green and Kolesar [27].

9.4.2 Response

OR research is quite active in the field of response management. It is only in recent years, in light of the growing terrorist threats and natural disasters, that the OR community has devoted significant attention to disaster management and humanitarian logistics. This research stream is continuously growing with the positive contribution that application of OR techniques can have on alleviating the consequence of an attack, driving the efficient planning of scarce resources [18]. Most OR research streams mainly concentrate on locating and dispatching ambulances, emergency vehicles and/or medical/emergency teams in the aftermath of a disaster to deal with the distribution of supplies and the medical treatment/transportation of injured people. However, research on transportation problems and vehicle fleet management for disaster response operations has only recently started emerging. In Luis et al. [42] and Pedraza-Martinez and Van Wassenhove [50] the reader can find interesting OR literature reviews on humanitarian logistics and routing problems for disaster response.

It should be mentioned that regardless of whether the cause of a disaster is natural (e. g., earthquakes, floods, hurricanes) or manmade (e. g., terrorist attacks, sabotage), its consequence might trigger large-scale loss of life as well as considerable damage to public/private infrastructure, houses and industrial complexes. As mentioned in the previous paragraphs, the impact of a disaster can significantly be influenced by the efficacy and the effectiveness of the logistics operations in place during the response phase. Although several casualties are directly produced by the disaster, a large fraction of the affected persons usually die in the aftermath of a disaster due to a lack of medical aid [6, 31].

Several features of the disaster triggered by intentional actions are to be considered while planning an emergency response. For example, the concentration of scarce resources (e. g., ambulances, medical resources) in the aftermath of a mass casualty incident is a reasonable assumption within traditional emergency response plans to allow a better coordination of the emergency response, increasing the visibility on the persons affected by the incident itself and optimizing the use of scarce resources [16]. However, in a mass casualty incident resulting from a terrorist act, having those resources concentrated in a specific area, might create a tempting high payoff target for a secondary attack. Nevertheless, dispersing resources would require additional communications and coordination efforts to deploy an effective and efficient emergency response. In addition, the decision making process during a disaster response differs from conventional decision making. In fact, while dealing with a disaster emergency response many variables of the problem are uncertain, with much of the information simply not available or, even if available, most often not reliable. The availability of information in such situations is somehow dependent on the lifecycle of a disaster, with information usually limited at the beginning, but ever more accurate as time passes. In addition, the environment and many of the problem features might rapidly change, and some critical disaster response decisions might be irreversible [49].

Despite all these challenges, all response actions to a disaster are to be executed under extremely challenging conditions such as limited availability of resources (transportation, supplies, manpower, hospital capacity), damaged transportation and communication infrastructure, as well as uncertain information regarding the number and locations of people in need of medical assistance [47, 74]. Moreover, it is crucial that the logistics relief operations are effectively planned, initiated as quickly as possible and efficiently coordinated. Therefore, there is a strong need for OR tools, which can support decision making, solving complex problems in limited time [6]. The use of optimization modeling to tackle emergency logistics has been growing continuously in the past decades. Caunhye et al. [13] reviewed the optimization models used in emergency logistics. Disaster operations are classified based on two possible time horizons: operations performed before or after disaster occurrence. *Pre-disaster operations* include OR problems coping with short-notice evacuation, emergency facilities location and stock pre-positioning. *Post-disaster operations* cover the processes of relief distribution and casualty transportation.

In post-disaster situations one of the critical factors influencing the delivery relief goods (e. g., food, shelter, medical supplies) to affected regions is the state of the road network. In many cases, it is not a lack of supplies that kills people, but the impossibility to get those supplies to the people who need them (Maya-Duque et al. [17]). Several authors have modeled the problem of transporting vital first-aid commodities and emergency personnel to disaster affected areas. Since ambulances often represent a scarce resource in disaster situations, their efficient usage is of the utmost importance. In Talarico et al. [62] a decision support approach for the routing of ambulances in response to a disaster scenario is described. In this case, a large number of injured persons requires medical aid at the same time, the number of ambulances, the medical personnel and the hospital capability being the limiting factors. Two types of patients are considered: 1) seriously injured patients who need to be brought to a hospital by an ambulance; 2) slightly injured persons who can be assisted directly in the field. Simulation is carried out to assess how the response to the emergency can vary depending on the number of ambulances, the number of hospitals, and the type of patients.

During post-disaster management a crucial activity is represented by assigning incoming emergency requests to ambulances, fire brigades, rescue teams and so on, depending on the type of request. All these assignment activities are sometimes classified as dispatching. In Toro-DíAz et al. [66] an integrated location and dispatching optimization model is proposed to assign locations and requests to the vehicles. A genetic algorithm is developed and a simulation is performed to test the impact of queuing patients in congested server systems on the achieved response time and coverage. A similar work is proposed by Andersson and Värbrand [2] whose model focuses mainly on the specific features of the problem of dispatching ambulances. Two main factors are used to support decisions: the urgency of requests and the closeness of a vehicle to the site of an incident.

Luis et al. [42] provide an extensive survey on OR applications on post-disaster relief, focusing on routing problems. In general, the main objective is to optimize the distribution of humanitarian aid supplies (e. g., water, food, medicine, and survival equipment) from distribution centers to demand points like refugee camps with respect to the available transport capacities. For example, in Campbell et al. [11] a variation of the traveling salesman and vehicle routing problems is proposed to minimize the latest arrival time at demand locations. In Wex et al. [71] an application of the multiple traveling salesman problem, which is one of most well-known problems in the OR domain, to plan the routing of rescue units used to serve a list of areas/point affected by a set of incidents is proposed. Some studies also propose a multi-commodity flow model combining the distribution of supplies with the transportation of patients. An example of such an application is described in Yi and Özdamar [75] in which a support decision model aims at minimizing the weighted sum of unsatisfied demands and waiting times of injured people.

9.4.3 Recovery

This area of research covers the planning of actions during the *reconstruction phase*, after a main incident. According to Altay and Green [1] this represents one of the main areas where more OR research is needed. In a survey on OR applications for disaster recovery, Kunz and Reiner [37] highlight the extreme importance of this stage since the quality of the logistic activities during this phase strongly impacts the success of the whole disaster recovery process, especially in terms of sustainability and long term effectiveness.

The majority of the existing OR work specialized in the recovery stage does not really differentiate between the industrial sector and/or the cause of the disaster. As mentioned in the Introduction, an attack targeting a critical chemical infrastructure might engender significant consequences on human life as well on the environment, also triggering partial and/or total disruptions of main logistics and utility networks.

Although the peculiarities of security-related incidents within the chemical and process industry might play a significant role in the associated decision making process, the typical decisions tackled by OR models in the recovery stage do not significantly differ from recovery decisions in the case of other safety-related accidents in the chemical sector and/or in other industrial domains. In fact, the typical decisions covered by OR models in this stage concern the prioritization of repair works to restore pre-incident conditions, while optimizing the use of scarce resources.

As mentioned in the previous paragraph, in the aftermath of mass casualties, a number of persons might suffer from a lack of clean water, food, shelter and adequate medical care. Mass care needs to be provided to evacuated human and animal populations, as well as significant efforts should focus on the rebuilding of damaged infrastructure such as communication, transportation, water and electricity networks. To allow a speedy and effective recovery from the accident an adequate logistics reply is, therefore, crucial. It is not surprising that the majority of the disaster relief effort consists of logistics [23].

As has already been seen in Figure 9.2, after a major supply chain destruction, recovery works could take place in two stages: 1) in a short term time horizon where the priority is to connect isolated areas and to provide a first disaster relief. In this case, repair works are done in a sort of contingency mode and might take few days. Due to the emergency situation the decision making is more time-constrained and recovery activities are more difficult to schedule. 2) In a medium long term period, repair works should lead to a complete restoration of the supply chain. This phase can take months or perhaps years. In Özdamar and Ertem [48], an extensive survey is presented that focuses on OR methods and solutions enabling both short term and long term recovery work to restore the performance of a supply chain.

In general, the decision process covering short term repair works to face the emergency is based on a decision maker's experience. The downside of this approach is that the relationship between emergency repair work and relief distribution from the system perspective are neglected. A quantitative global optimization approach for road network repair work is proposed by Yan and Shih [73]. Their model can support and improve decision making aiming at minimizing the length of the time required for both emergency roadway repair and relief distribution. Several operating constraints, linked to the planning of real-life emergency repair activities, the relief distribution routes and schedules within a limited time are considered. A heuristic algorithm is proposed to efficiently solve this problem, which is tested on a case study. A similar work is proposed by Feng and Wang [22], which focuses on road network repair activities that occur over the first 72 hours after the supply chain disruption. A multi-objective optimization model is developed aiming at assigning the repair tasks to the work teams. The conflicting objectives that are optimized concern the maximization of the performance of the emergency road repair activities, the maximization of the number of people that could potentially benefit from these repair works and, finally, the minimization of the risk potentially incurred by rescue teams while providing help to affected areas.

As major road network disruptions tends to disconnect remote population centers from the main supply hub, a decision model to restore accessibility to towns and villages by planning the emergency repair of a rural road network is proposed by Maya-Duque et al. [45]. The decision model proposed integrates different phases of the disaster management process as it considers the emergency recovery of the road network, while capturing the urgency of repairing a road to facilitate the relief distribution. This model could be used to support recovery decisions covering both the short term and the long term horizon. The authors develop both an exact algorithm and an efficient heuristic, which are tested on small and medium instances. Experimental tests provide important managerial insights as follows: a) On average 30% of damaged nodes are to be repaired to restore full accessibility to the entire network; b) on average, each subsequent repair of a damaged node follows the law of diminishing return.

The planning of recovery activities is not an easy task, especially when faced with several constraints such as the availability of limited recovery budgets and the limited availability of repair capacity, which is a typical of poor and/or remote regions of the world. Therefore, it is crucial that repair activities are planned and executed in the most efficient way possible, restoring accessibility of demand points (e. g., small villages, medical facilities, shelters, food distribution points) to main supply centers. Liberatore et al. [40] propose a hierarchical optimization model considering different objective functions associated with distribution of relief supplies in a post-disaster setting where damaged roads with reduced reliability are considered. The model also proposes a list of road segments that need repair efforts, whose costs do not pass a limited budget. A multi-objective model is proposed by Matisziw et al. [43] focusing on short term recovery works on a telecommunication network. To allow an effective and rapid restoration of the communication services in a damaged network, damaged components are to be repaired or reconfigured. These activities might be time consuming and costly. The proposed decision model allows a prioritization of the network repair activities maximizing the system performance over a planning horizon subject to budgetary restrictions. Two different objectives are considered: minimization of repair cost and the maximization of system flow. The approach is tested on an illustrative example of a telecommunication network.

An approach to provide full restoration of critical infrastructures such as essential public services (i. e., power, telecommunications, water and transportation) after a main disruption is described in Matisziw et al. [43]. The objective is to develop an integrated plan and support short and long term managerial decisions aimed at restoring essential interdependent public services that have been directly or indirectly spoiled by an accident. This is quite important since a timely restoration of these basic services is necessary for society to recover from a disaster. A mixed-integer program (MIP) that integrates three different types of decisions is developed. 1) Identification of components to be installed or repaired; 2), the assignment of work task on the selected components to available team; and 3) the order in which each work group will complete the tasks assigned to it. The objective function of this problem measures how well the services are being restored over the horizon of the restoration plan. A heuristic solution method is proposed and tested on a test case representing the power and telecommunication systems of a large portion of Manhattan.

9.5 Conclusions

In this chapter, some OR applications which can be applied in the chemical industry to increase the overall level of security have been presented. The models have been grouped into four different sets of applications (mitigation, preparedness, response and recovery), depending on the life cycle of an attack. Those categories group specific models and methods used to support the decision making process in each stage of an intentional attack. Those practical applications could definitely support decision makers increase their risk awareness and find appropriate solutions to deal with the security threats. Some OR applications in the field of cyber security are also mentioned. Those applications could help contrasting large-scale cyber attacks on the information infrastructure of a critical infrastructure.

It should be clearly stated that a terrorist attack will pose a serious security threat also in the future, with significant implications in modern communities, businesses and economies. Therefore, it is truly important to better understand how to cope with such security threats in an effective and efficient manner. Indeed, better risk awareness could definitely lead to a better management of disaster operations, improving readiness as well as increasing the speed of response and recovery.

The use of OR could definitely support planning and control activities in order to improve the emergency response in the chemical industry as well as enhance the decision making process to define more effective preventive actions. In spite of all the new technologies and OR methods to improve security, all threats cannot be completely removed. A residual risk will always remain. Therefore, a more risk aware security management will definitely improve the level of preparedness. This requires the application of robust and effective early warning systems and adoption of effective contingency plans. Continuous research and development is necessary to address changing threats and related challenges.

Innovative technologies might offer significant contributions in the security field. For example, evolved families of sensors might offer significant capabilities to improve security, by increasing early warning detection and reducing nuisance. Together with the evolution of technology, the integrated form of communications might support the development and the application of innovative security solutions.

Security related OR research will continue to be an important enabling activity in the future. It will enable more efficient and effective decision making to prevent, neutralize and/or mitigate the consequence of intentional acts on critical infrastructures, process, systems and/or any other viable targets. The increasing need for interdisciplinary research will continuously encourage the transfer or existing OR knowledge in the security field. This will enable more future research challenges in the OR field aimed at developing new and more accurate mathematical models to consider all complex interconnections typical of the chemical and process industry in order to respond effectively to large-scale intentional disasters.

The literature on OR applications to increase the security level in the chemical sector is continuously growing, but many issues still deserve further exploration such as cyber security, critical infrastructure protection, threat analysis, risk analysis and so on. In addition, OR could also find new research paths exploring the application of OR models and techniques to predict the likelihood of potential attacks by combining traditional OR methodologies and data mining applications.

References

[1] Altay N, Green WG. (2006). OR/MS research in disaster operations management. Eur J Oper Res. 2006;175(1):475–493.

[2] Andersson T, Värbrand P. Decision support tools for ambulance dispatch and relocation. Oper Res Soc. 2007;58(2):195–201.

[3] Bastien MC, Dumas M, Laporte J, Parmentier N. Evacuation risks: a tentative approach for quantification. Risk Analysis. 1985;5(1):53–61.

[4] Batta R, Kwon C. Handbook of OR/MS models in hazardous materials transportation. Berlin: Springer; 2013.

[5] Bell JE, Griffis, Stanley E, WAC, Eberlan JA. Location optimization of strategic alert sites for homeland defense. 2011;Omega, 39(2):151–158. http://doi.org/http://dx.doi.org/10.1016/j.omega.2010.05.004

[6] Berkoune D, Renaud J, Rekik M, Ruiz A. Transportation in disaster response operations. Soc Econ Plann Sci. 2012;46(1):23–32.

[7] Branscomb LM, Klausner RD et al. Making the nation safer: the role of science and technology in countering terrorism. Committee on Science and Technology for Countering Terrorism, National Research Council; 2002

[8] Brown G, Carlyle M, Salmerón J, Wood K. Defending critical infrastructure. Interfaces 2006;36(6):530–544.

[9] Bula GA, Prodhon C, Gonzalez FA, Afsar HM, Velasco N. Variable neighborhood search to solve the vehicle routing problem for hazardous materials transportation. J Hazard Mat. 2017;324:472–480.

[10] Calvo RW, Cordone R. A heuristic approach to the overnight security service problem. Comput Oper Res. 2003;30(9):1269–1287.

[11] Campbell AM, Vandenbussche D, Hermann W. Routing for relief efforts. Transport Sci. 2008;42(2):127–145.

[12] Carmichael F. A guide to game theory. London: Pearson Education; 2005.

[13] Caunhye AM, Nie X, Pokharel S. (2012). Optimization models in emergency logistics: A literature review. Socio Econ Plan Sci. 2012;46(1):4–13. http://doi.org/http://dx.doi.org/10.1016/j.seps.2011.04.004

[14] Current J, Min H, Schilling D. Multiobjective analysis of facility location decisions. Eur J Oper Res. 1991;49(3):295–307. https://doi.org/10.1016/0377-2217(90)90401-V.

[15] De Keyser W, Springael J. Why don't we kiss!?: A contribution to close the gap Between real-world decision Makers and theoretical decision-model builders. ASP/VUBPRESS/UPA; 2010.

[16] der Heide EA. The importance of evidence-based disaster planning. Ann Emerg Med. 2006;47(1):34–49.

[17] Duque PAM, Coene S, Goos P, Sörensen K, Spieksma F. The accessibility arc upgrading problem. Eur J Oper Res. 2013;224(3):458–465. https://doi.org/10.1016/j.ejor.2012.09.005.

[18] Ergun O, Karakus G, Keskinocak P, Swann J, Villarreal M. Operations research to improve disaster supply chain management. Wiley Encyclopedia of Operations Research and Management Science; 2010.
[19] EURO. What is operational research? Retrieved from https://www.euro-online.org/web/pages/301/or-and-euro; 2017.
[20] Fairley WB. Market risk assessment of catastrophic risks. In HC Kunreuther, Ley EV (editors.) The risk analysis controversy: An institutional perspective, pp. 195–199. Berlin, Heidelberg: Springer; 1982. http://doi.org/10.1007/978–3-642–81940-7_14
[21] Farahani RZ, Asgari N, Heidari N, Hosseininia M, Goh M. Covering problems in facility location: A review. Comput Ind Eng. 2012;62(1):368–407.
[22] Feng C-M, Wang T-C. Highway emergency rehabilitation scheduling in post-earthquake 72 hours. Journal of the 5th Eastern Asia Society for Transportation Studies. 2003;5(3281), 3276–3285.
[23] Glenn Richey Jr R, Kovács G, Spens K. Identifying challenges in humanitarian logistics. Int J Phys Distrib Logist Manag. 2009;39(6):506–528.
[24] Glickman TS, Rosenfield DB. Risks of catastrophic derailments involving the release of hazardous materials. Manage Sci. 1984;30(4):503–511.
[25] Golalikhani M, Karwan MH. The effect of weather systems in hazmat transportation modeling. In Handbook of OR/MS models in hazardous materials transportation. Berlin: Springer; 2013. pp. 103–125
[26] Gopalan R, Kolluri KS, Batta R, Karwan MH. Modeling equity of risk in the transportation of hazardous materials. Oper Res. 1990;38(6):961–973. Retrieved from http://www.jstor.org/stable/170964.
[27] Green LV, Kolesar PJ. Anniversary article: Improving emergency responsiveness with management science. Manage Sci. 2004;50(8):1001–1014.
[28] Grossi P. Catastrophe modeling: A new approach to managing risk, Vol 25. Springer Science and Business Media; 2005.
[29] Hennessy JL, Patterson DA, Lin HS. Information technology for counterterrorism: Immediate actions and future possibilities. Committee on the role of information technology in responding to terrorism. The National Academies Press, National Research Council, Washington, DC; 2003.
[30] Hernandez M. Mexican rebels claim pipeline attacks. The Washington Post; 2007.
[31] Holguín-Veras J, Jaller M, Van Wassenhove LN, Pérez N, Wachtendorf T. On the unique features of post-disaster humanitarian logistics. J Oper Manage. 2012;30(7):494–506.
[32] INFORM. What is operations research? Retrieved from https://www.informs.org/About-INFORMS/What-is-Operations-Research; 2017.
[33] Janssens J, Talarico L, Sörensen K. A hybridised variable neighbourhood tabu search heuristic to increase security in a utility network. Reliab Eng Syst Safe. 2016;145:221–230.
[34] Jerman-Blažič B, et al. An economic modelling approach to information security risk management. Int J Inform Manage. 2008;28(5):413–422.
[35] Jia H, Ordóñez F, Dessouky M. A modeling framework for facility location of medical services for large-scale emergencies. IIE Trans. 2007;39(1):41–55. http://doi.org/10.1080/07408170500539113.
[36] Kaplan E. Targets for terrorists: Chemical facilities. Council on Foreign Relations; 2006.
[37] Kunz N, Reiner G. (2012). A meta-analysis of humanitarian logistics research. J Humanitarian Logistics and Supply Chain Management. 2(2):116–147.
[38] Lancaster University Management School. What is operational research? Retrieved February 1, 2017 from http://www.lancaster.ac.uk/lums/study/masters/programmes/msc-operational-research-management-science/what-is-operational-research/.
[39] Larson RC, Metzger MD, Cahn MF. Responding to emergencies: Lessons learned and the need for analysis. Interfaces. 2006;36(6):486–501.

[40] Liberatore F, Ortuño MT, Tirado G, Vitoriano B, Scaparra MP. A hierarchical compromise model for the joint optimization of recovery operations and distribution of emergency goods in Humanitarian Logistics. Comput Oper Res. 2014;42:3–13.
[41] Luft G. Pipeline sabotage is terrorist's weapon of choice. Pipeline Gas J. 2005;232(2):42–44.
[42] Luis E, Dolinskaya IS, Smilowitz KR. Disaster relief routing: Integrating research and practice. Soc Econ Plann Sci. 2012; 46(1):88–97.
[43] Matisziw TC, Murray AT, Grubesic TH. Strategic network restoration. Netw Spat Econ. 2010;10(3):345–361.
[44] May PJ, Jochim AE, Sapotichne J. Constructing homeland security: An anemic policy regime. Policy Stud J. 2011;39(2):285–307.
[45] Maya-Duque PA, Dolinskaya IS, Sörensen K. Network repair crew scheduling and routing for emergency relief distribution problem. Eur J Oper Res. 2016; 248(1):272–285.
[46] Michallet J, Prins C, Amodeo L, Yalaoui F, Vitry G. Multi-start iterated local search for the periodic vehicle routing problem with time windows and time spread constraints on services. Compu Oper Res. 2014;41:196–207.
[47] Najafi M, Eshghi K, Dullaert W. A multi-objective robust optimization model for logistics planning in the earthquake response phase. Transport Res E-Log. 2013;49(1):217–249.
[48] Özdamar L, Ertem MA. Models, solutions and enabling technologies in humanitarian logistics. Eur J Oper Res. 2015;244(1):55–65.
[49] Pauwels N, Van De Walle B, Hardeman F, Soudan K. The implications of irreversibility in emergency response decisions. Theor Decis. 2000;49(1):25–51.
[50] Pedraza-Martinez AJ, Van Wassenhove LN. Transportation and vehicle fleet management in humanitarian logistics: challenges for future research. EURO Journal on Transportation and Logistics. 2012; 1(1–2), 85–196.
[51] Poole-Robb S, Bailey A. Risky Business: Corruption, fraud, terrorism, other threats to global business. Brigham Young University International Law and Management Review. 2007;3(1):141–142.
[52] Reniers GLL, Ale BJM, Dullaert W, Foubert B. Decision support systems for major accident prevention in the chemical process industry: A developers' survey. J Loss Prevent Process Ind. 2006;19(6):604–620. http://doi.org/10.1016/j.jlp.2006.02.005.
[53] Reniers GLL, Sörensen K, Dullaert W. A multi-attribute systemic risk index for comparing and prioritizing chemical industrial areas. Reliab Eng Syst Safe. 2012;98(1):35–42.
[54] Reniers GLL, Sörensen K, Khan F, Amyotte P. Resilience of chemical industrial areas through attenuation-based security. Reliab Eng Syst Safe. 2014;131, 94–101. http://doi.org/http://dx.doi.org/10.1016/j.ress.2014.05.005.
[55] Scanlon RD, Cantilli E. Assessing the risk and safety in the transportation of hazardous materials. 1985
[56] Scott A. Terrorist attack hits U. S.-owned chemical plant in France. Chem Eng News; 2015 Retrieved from http://cen.acs.org/articles/93/web/2015/06/Terrorist-Attack-Hits-US-Owned.html
[57] Sheffi Y, Rice Jr JB. A supply chain view of the resilient enterprise. MIT Sloan Management Review. 2005;47(1):41.
[58] Sörensen K. Metaheuristics – the metaphor exposed. Int T Oper Res. 2015;22(1):3–18.
[59] Srivastava A, Gupta JP. New methodologies for security risk assessment of oil and gas industry. Process Saf Environ. 2010;88(6):407–412.
[60] Starčević SM, Gošić AM. Methodology for choosing a route for transport of dangerous goods: Case study. Vojnotehni{č}ki Glasnik, 2014;62(3):165–184.
[61] Taha H A. Operations research: an introduction, Vol 557. Pearson/Prentice Hall; 2007.
[62] Talarico L, Meisel F, Sörensen K. (2015). Ambulance routing for disaster response with patient groups. Comput Oper Res, 56, 120–133. http://doi.org/10.1016/j.cor.2014.11.006

[63] Talarico L, Reniers G, Sörensen K, Springael J. MISTRAL: A game-theoretical model to allocate security measures in a multi-modal chemical transportation network with adaptive adversaries. Reliab Eng Syst Safe. 2015;138:105–114. http://doi.org/10.1016/j.ress.2015.01.022

[64] Talarico L, Sörensen K, Reniers G, Springael J. Pipeline security. Securing Transportation Systems. 2015;281–311.

[65] Talarico L, Sörensen K, Springael J. The k-dissimilar vehicle routing problem. Eur J Oper Res. 2015;244(1):129–140. http://doi.org/10.1016/j.ejor.2015.01.019.

[66] Toro-Díaz H, Mayorga ME, Chanta S, Mclay LA. Joint location and dispatching decisions for emergency medical services. Comput Ind Eng. 2013;64(4):917–928.

[67] Tufekci S, Wallace WA. The emerging area of emergency management and engineering. IEEE T Eng Manage. 1998;45(2), 103–105.

[68] Van Raemdonck K, Macharis C, Mairesse O. Risk analysis system for the transport of hazardous materials. J Safety Res.2013;45:55–63.

[69] Viduto V, Maple C, Huang W, López-Peréz D. A novel risk assessment and optimisation model for a multi-objective network security countermeasure selection problem. Decis Support Syst 2012;53(3):599–610.

[70] Wein LM, Baveja M. Using fingerprint image quality to improve the identification performance of the US Visitor and Immigrant Status Indicator Technology Program. Proceedings National Academy of Sciences of the United States of America. 2005;102(21):7772–7775.

[71] Wex F, Schryen G, Feuerriegel S, Neumann D. Emergency response in natural disaster management: Allocation and scheduling of rescue units. Eur J Oper Res. 2014;235(3):697–708.

[72] Wright PD, Liberatore MJ, Nydick RL. A survey of operations research models and applications in homeland security. Interfaces. 2006 36(6):514–529.

[73] Yan S, Shih Y-L. Optimal scheduling of emergency roadway repair and subsequent relief distribution. Comput Oper Res. 2009;36(6):2049–2065.

[74] Yi P, George SK, Paul JA, Lin L. Hospital capacity planning for disaster emergency management. Soc Econ Plann Sci. 2010;44(3), 151–160.

[75] Yi W, Özdamar L. A dynamic logistics coordination model for evacuation and support in disaster response activities. Eur J Oper Res. 2007;179(3):1177–1193.

[76] Zamparini L. Transport security in EU and US: Competing or complementary visions. In Nectar Workshop on Transport Security, Lecce, Italy, pp. 5–6; 2010.

[77] Zhang L, Reniers G. A game-theoretical model to improve process plant protection from terrorist attacks. Risk Analysis. 2016

10 Conclusions

Many warning signals indicate that terrorism is evolving in the direction of an ever larger attempt to cause mass casualties. Nonetheless, physical security within chemical plants against potential terrorist attacks is still largely insufficient and predominantly compliance driven. Furthermore, observations indicate that safety legislation in the chemical industry has mainly been issued ad hoc, as a political reaction to major accidents such as Flixborough (1974), Seveso (1976), Bhopal (1984), Mexico City (1984), Basel (1986), Chernobyl (1986), Piper Alpha (1988), Enschede (2000), Toulouse (2001), Texas City (2005), Buncefield (2005), Deepwater Horizon (2010), and others, and regretfully, it seems that in Europe security legislation follows the same trend. In the United States, there is centralized and top down anti-terrorism legislation for the process industry. In Europe, existing legislation on anti-terrorism security is scattered and bottom-up. The result is that security is only taken seriously in those European facilities where top-management is convinced about its usefulness. Otherwise, anti-terrorism policy and security measures are very often simply absent in European chemical plants.

Thus far, luckily no major terrorist attack has been successfully carried out on a chemical facility in the western world. However, it is not because an event has not happened, that it will never happen. Chemical security should not be taken for granted. One major successful terrorist attack on a chemical plant would be enough to fundamentally change the way that western societies deal with chemical security, and the way these societies look at chemical production and storage.

Safety at chemical facilities starts with safety risk assessments, and a knowledge of all safety risks present. Security in chemical industrial areas is similar: all security risks need to be known, otherwise security cannot be managed. The most dangerous risks, whether they are unintentional (safety-related) or intentional (security-related), are those that are not known. Only those security risks that have been identified, analyzed and prioritized, can be actively managed, and countermeasures can be taken for those known risks. Thus, adequate security starts with reliable and valid security risk assessments. This book provides a state-of-the-art overview of what security risk assessments should be composed of, and advances current knowledge by discussing current research and also giving indication towards what future research should be concerned with on this matter.

Besides the differences between European and US legislation on antiterrorism security and the topic of what security risk assessments should be about and how they can be improved, another chapter in this book is concerned with the use of methods from operations research to further optimize security with respect to the chemical process industry.

Academics, policy makers and captains of industry should realize that research and development with respect to security, proactive as well as reactive, is urgently needed in European chemical industrial areas. Only by such research and development efforts, followed by implementation, can antiterrorism security be truly ensured in the chemical industry.

https://doi.org/10.1515/9783110499087-010

Conclusions

Index

https://doi.org/10.1515/9783110499087-011

Also of interest

Energetic Materials Encyclopedia.
Klapötke, 2018
ISBN 978-3-11-044139-0, e-ISBN 978-3-11-044292-2

Engineering Risk Management.
Meyer, Reniers, 2016
ISBN 978-3-11-041803-3, e-ISBN 978-3-11-041804-0

Research Laboratory Safety.
Kuespert, 2106
ISBN 978-3-11-044439-1, e-ISBN 978-3-11-044443-8

Process Technology.
An Introduction
De Haan, 2015
ISBN 978-3-11-033671-9, e-ISBN 978-3-11-033672-6

Bei Fragen zur Produktsicherheit wenden Sie sich bitte an:
If you have any questions regarding product safety,
please contact:

Walter de Gruyter GmbH
Genthiner Straße 13
10785 Berlin
productsafety@degruyterbrill.com